PRESERVED
STEAM TRACTION

PRESERVED STEAM TRACTION

ERIC SAWFORD

 Patrick Stephens, Wellingborough

Half-title page An attractive view of three wagons which have made the long journey to Australia and eventually returned to Britain to be carefully restored to their former glory. From left to right Aveling and Porter 9282 of 1921, Atkinson six-ton 72 of 1918 and Foden six-ton Colonial 4086 of 1913. The Foden is no longer in this country.

Title page A quiet moment for the crew of a Fowler Class 'BB' ploughing engine, 14383 *Prince*, seen here with a reversible four-furrow plough. The drum on the engine would carry 1,000 yards of rope. Many items of equipment have been built to enable these powerful engines to plough and cultivate and also to carry out dredging of ponds and lakes.

First published in 1985

British Library Cataloguing in Publication Data

Sawford, E.H.
  Preserved steam traction.
  1. Traction-engines—Great Britain—Pictorial works
  I. Title
  629.2'292'0941   TJ700

ISBN 0-85059-739-0

Patrick Stephens Limited is part of the Thorsons Publishing Group.

Photoset in 10 on 10 pt Plantin by Avocet Marketing Services Limited, Aylesbury, Bucks. Printed in Great Britain on 115 gsm Fineblade coated cartridge, and bound, by The Garden City Press, Letchworth, Herts, for the publishers, Patrick Stephens Limited, Denington Estate, Wellingborough, Northants, NN8 2QD, England.

Contents

A front view of the rare Robey six-ton wagon, thought to be the
only Robey wagon in the country. Seen here on one of its very
occasional outings, the 1973 Carrington Park rally held in
Lincolnshire.

Introduction by
Dr J. L. Middlemiss

The study of steam road vehicles has become a subject in its own right. Such vehicles are viewed now with more interest than they were 50 or 60 years ago. Then they were viewed very differently — as a means of making a person's, or a firm's living, and taken very much for granted.

In the engine 'movement' today, all walks of life are represented, the professions, model engineers, artists, statisticians, photographers — many are people with no previous steam experience. The author of this book, a photographer amongst other things, has fortunately been a steam road vehicles' enthusiast for many years.

Today, engines are being renovated and rebuilt, often under very restricted and limiting conditions, requiring years of labour. But it is done — for example, great weights can be lifted without a crane using a 20-ton jack and plenty of timber. Rivetting can be done by three people, using a hand-blown coke forge.

The writer of this foreword obtained his engine (which he has known as long as he can remember) in a derelict state, over 30 years ago. She was lying derelict on a farm in Northumberland. New-laid eggs were near the firedoor, and cats flew out from underneath.

A fire was lit, and the engine steamed. Knowing what we do now, we took a chance, and the smokebox was so thin that earth was put in to make it airtight. When it was renewed it was thin enough to cut my son's hand. Yet it had held up the whole front end of the engine. Nowadays, engines are advertised, and high prices obtained. But many years ago, only word of mouth pinpointed an engine, often lying down a lane, where it could be bought at scrap price. The engine (often red rust) was towed away by Land Rover.

In 1951, two showman's engines were abandoned on the banks of Loch Lomond and they were eventually scrapped. No engines are being scrapped now. Some amazing renovations have been accomplished. A large number have had their worn-out fireboxes removed, and replaced, a remarkable feat when one considers that until fairly recently, a worn-out box was adequate reason for cutting up an engine. Unusual engines, and many unique examples have been brought back from overseas to this country, particularly Australia (where many firms had good agents) Mozambique and the USA. It is quite obvious that the engines are going to outlive their owners. Steam struggled on commercially till the early 1960s, an old and experienced employee being kept on to drive a steam wagon or roller. When he retired, an enthusiast might put forward a tender to the council, or the employer, to buy it.

The first rally occurred at Appleford, on June 24 1951 beginning as a friendly wager, for a firkin of beer following a race. In those days no appearance money was paid and engine racing was the norm. Engines travelled to and from events on their own steam, low loaders being seldom used. Of course, the ordinary traffic was not as dense as it is now.

Rallies have become now part of the summer scene in this country. Most areas have their steam clubs. Some rallies have now passed their Silver Jubilees, and fix their date in the calendar so as not to clash with others. Other creations of the mechanical world are included, and articles bought and sold. Appearance money is paid and coal provided. In other words, whether we like it or not, economics have crept in.

For many years my engine did the 100-mile return journey to a Durham rally. Engines sometimes ran out of coal on the way, or had to complete the trek after 'lighting up time'. A 20 to 30 mile journey was nothing unusual. A rally would usually occur on a Saturday afternoon, not over a whole weekend, and a big rally might comprise 12 engines.

Today the procedure to take an engine on the road is surprisingly complicated. To begin with, the boiler must have an annual inspection. On receiving a satisfactory report from the boiler insurance company's inspector (£50), the insurance company covering road work and appearance at public events, will then grant a certificate (£30). This certificate in turn is transmitted to the road tax authority. The road fund licence (£15) permits the engine, usually called an agricultural machine, to use any road, but its use is restricted, and it may tow nothing other than for the engine and its occupants, for example a coal and water trailer and a living van.

There is no tax payable for a diesel or steam roller, as that machine is doing a common good in rolling the Queen's highway. There have been a number of unpleasant accidents in recent years on busy roads and most engine drivers of experience prefer quieter routes. Engines are not allowed on motorways. There are other problems too. Water points are becoming fewer in number, and an agricultural engine needs water every ten miles or so. A road locomotive or showman's engine can cover 20 to 25 miles on one tank but a roller needs water every five to six miles. Coal requirements average about one 1 cwt bag per hour.

One is often asked how fast a machine will travel. If a three-speed road engine on rubbers averages five to six miles per hour in a day's travelling she is doing well. Rollers may average two miles per hour, an agricultural engine on steel wheels no more than three miles in an hour. If a roller has done 25 miles in a day she has done well, and so has her driver. The noise of the gears, and the resonance of the rollers transmitted up to the canopy and down to the driver has to be experienced to be believed.

There are very few engines now which still carry their original livery and paintwork, although the colour schemes and lining-out of most of the makers are well known. Restraint is necessary and the temptation to over-decorate must be strenuously resisted. The dynamo on a showman's engine was to light the fair, and if there were a few amperes left over, a few lamps were fitted near the footplate and in the engine's canopy as an afterthought. It is also possible to over-decorate an engine with brasswork (all the more to polish) and in the early days, engines were taken to rallies in a much more run down state than now. Some boilers were regarded as suspect and dripped water from places where they should not have leaked, and were avoided by those who knew better. Gears are renewed now, and their teeth built up, bearing brasses are built up or renewed and, in fact, some are better now than when they left the place of their birth. Car showroom finish, chrome-plated coupling pins and rope rollers are other unnatural additions.

Many an old timer had never thought to see an engine in steam again, (let alone several in a row) and it could be quite touching to see some lay an affectionate hand on an engine, eyes gleaming, breathing deeply the oil-steamy air, as they listened again to the gentle chuffing of the favourite engine of their youth. On and around the engines the old brigade would maintain a constant flow of banter, reminiscences and argument. Many an adventure would be relived, many a harvest re-threshed, and the rival virtues of Burrells, Garretts and Marshalls became, again, matters of moment instead of history. We of the middle generation could only listen in awe, regretting the lateness of our birth.

This phenomenon has now gone full circle, but there are still a number of engine owners who are fortunate enough to remember engines when they were earning their living.

Dr J.L.Middlemiss
March 1985

Preface

Within the photographs selected for this book can be found a record of nearly 25 years of the traction engine preservation scene. This book covers only engines capable of moving under their own power. Other steam exhibits, such as the portable engine, centre engines from fairground rides, fire pumps and stationary engines also exist but these are a subject in their own right and so have not been included here.

Engine rallies in the 1960s were considerably different from those of today. In the early days vast crowds of people were not the order of the day, many of the engines attending would be in the process of restoration and the vast majority of engines would be travelling under their own steam. Events of today see many engines in superb condition, some possibly even better than when new. In addition a great many other attractions have been added to interest the considerable numbers of people who now attend rallies. The present day engine scene would be incomprehensible to the long departed engineers who built these fine examples of British craftmanship. A rally programme from one of the larger events can show the considerable number and types of engines that survive. There are over two and a half thousand engines in the British Isles and the number is still slowly rising, as examples of the British builders' products exported to distant parts many years ago return home from time to time. On rare occasions nowadays an indigenous engine is discovered.

At most events at least one example of the magnificent showman's engine can be found, while at some of the major events the number attending can easily reach double figures. These engines are mainly the products of three major builders, Burrells, Fowlers and Fosters. They present a truly fascinating sight with their twisted brasswork and the dynamo mounted on an extension of the smokebox, providing power for both the ride and the coloured lights. Stand awhile near one of these superb engines as it drives a ride, or powers an organ, and listen carefully to the enthralling sounds that it makes as it provides the power — a truly memorable sound, accompanied by the smell of hot oil and steam. Just imagine the larger fairs of yesteryear, when several of these engines in their working days, could be found in and amongst the fairground rides. After the fair closed the wagon train would be loaded up, the showman's road locomotive would be at its head, soon to get to grips with its load on its way to the next fair. These engines were to be seen at the head of several wagons and given a reasonable road and conditions, could make good time to the next location.

A close relative of the 'showman's are the heavy haulage locomotives which are mostly between 8 and 10 NHP, three-speed engines. These were to be found moving the extremely heavy or difficult loads. Nowadays such loads can be seen occasionally behind modern and extremely powerful diesel heavy haulage units. Carefully restored examples of the heavy haulage locomotives can be seen. While not carrying the trimmings of the showman's, they are every bit as powerful and prove fascinating subjects for observation and camera alike. Haulage engines exist in various sizes down to the nippy little tractors built by most of the companies. These units are easily handled, speedy, and were used in their day for lighter delivery and haulage jobs.

The types which outnumber all others in preservation, are the general purpose traction engines and the road rollers. The general purpose agricultural engine was the 'maid of all work' around the farm, its duties consisting of threshing, wood sawing, driving crushing plants and of course, haulage on the farm. Many of those we see today have been carefully restored after many years of lying dumped and slowly rotting away in some isolated part. This type of engine was one of the 'bread and butter' lines of most engine builders, from the largest, through to the smallest, including those who did not build many engines.

Many people can remember the steam roller at work, mostly accompanied by its living van and water cart, forming its own train when travelling to and from location. The Aveling company monopolised the roller market and outsold all others. As the roller was still in use by councils and contractors up until the 1960s, many have survived. The engine preservation movement was well under way when many of the rollers were replaced by the more modern equipment — as a result they were sought after and purchased by steam enthusiasts. Many steam rollers built by British companies found themselves thousands of miles away and carried out the work for which they were built in conditions unlike anything in this country. Several much-travelled rollers have returned to Britain in recent years.

Steam wagons also provide a fascinating subject and examples of the 'undertype' with the engine mounted under the body and fitted with a vertical boiler (mostly Sentinels) and the 'overtype', where the engine is on top of a horizontal boiler, exist in considerable numbers. Overtypes are mainly Fodens but examples built by other manufacturers also exist. Some of the wagons we see today have extremely spartan conditions for the crew, offering little, or no, protection against the elements. The steam wagon was the forerunner of today's diesel lorry and many of the later units have pneumatic tyres, electric lighting, wipers etc, and are capable of considerable speeds, even when fully loaded.

The last major type of engine includes the very impressive and powerful ploughing engines. Many exist today and most of them are Fowlers, who, without doubt, were the company who most specialised in steam ploughing engines and equipment. The famous double-ploughing system was in use for over a hundred years. It is still occasionally used to clean out lakes and ponds. The Fowler company manufactured a range of ploughs, cultivators, mole drainers etc. The equipment was used by agricultural contractors, many operating considerable numbers of ploughing sets, although the larger farmers might also own a number of sets.

Only two decades ago, the writer can recall eight Fowler ploughing engines lying in various states of dereliction in three separate locations within a few miles of Huntingdon. Like the steam roller, which was the last type of engine in regular use, the ploughing engines, due to their size were amongst the last types to be recovered for preservation.

Various other engine types can also be seen. Crane engines and timber tractors add variety to the already interesting selection of steam operated equipment with us today.

The photographs in this book have all been taken by the author over the years in many different locations. The brief descriptions of the engine builders are not intended to be a history of the firms, or their products. These have mostly been covered by others. The main intention is to briefly outline the products and to describe what has survived into preservation today. Other companies building traction engines did exist at various times, but either no known examples of their products survive, or the only examples still with us are portable engines, which are not included in this volume.

1 The Aveling Group

In the preservation movement more products of this well known manufacturer have been restored than any other. The Aveling engines and rollers are known and respected worldwide and examples of the company's products can still be found in a great many countries. Many people will associate the Aveling name with steam road rollers and they have of course built a great many over the years, but they also constructed agricultural and road locomotives, portables, tractors and wagons.

One of the pioneers of steam traction, Thomas Aveling was born in the county of Cambridgeshire in 1824. When the family moved to Rochester in Kent, Thomas Aveling became involved with farming. In 1850 he opened a small engineering shop in Rochester and his great interest in steam power soon became evident when he converted a Clayton and Shuttleworth portable to become self propelled. His first engine was in fact built by Clayton and Shuttleworth due to the limitations of his own resources.

The first really successful steam road roller was to make its appearance in 1867 after various developments on the experimental engine proved to be a success. Early rollers were mostly 12 and 15-ton forerunners of the many hundreds to be built by the company over the years.

It is not surprising in view of the numbers built that the highest number of Avelings to survive are road rollers, representatives of both single-cylinder and compound engines in several weights and designs. It would be difficult to attend an engine event without at least one Aveling roller being included in the engine line up.

As the steam roller was still actively engaged on the duties for which it was built until fairly recently, a great many were purchased by enthusiasts for preservation and many were in reasonable condition having been replaced by the modern diesel roller only in the 1950s and 1960s.

Several unique and extremely interesting engines built by Avelings are still with us. Firstly mention must be made of the only existing Aveling and Porter showman's engine. Carrying works number 4885, *Samson* was built in 1901 for the Admiralty, and as a result was fitted with several non-standard items. The engine has two water columns and two pressure gauges. When used by the Admiralty at HM Dockyard Chatham, a capstan was fitted to the rear axle, outside of a rear wheel, but it was later removed. The engine passed to an owner in Stowmarket, and was later converted to a showman's for Charles Prestland, being used by him until 1946, when it was purchased and used on agricultural duties, finally ending up once again in the county of Suffolk. In 1965 the engine came to the notice of a Mr Paisley who acquired it and began restoration immediately, the engine being steamed at the private events held occasionally at Holywell and at a few local steam events, including the large gathering at 'Expo' in 1976. The engine was sold in the Holywell sale held on October 15 1980, when it became part of the Gisburn collection. *Samson* is an 8 NHP double-crank compound two-speed engine.

Another sole survivor is the Aveling five-ton wagon number 9282 of 1921. This wagon was exported to a location near Sydney, Australia. After some years it was sold to Mudgee council for whom it worked until 1936, ending up at a gold sluicing plant for a few years before it was laid aside and became derelict. In 1978 it was shipped back to Britain, where it was restored in a matter of months and it is now thought to be the only Aveling wagon in this country.

In order to work singlehanded within the heavy motor car regulations Avelings designed a speedy 4 NHP two-speed tractor, with improved springs, large tanks and with a winding drum and steel rope. Other road locomotives were frequently built in several sizes. The company also built ploughing engines and two survive at Thursford. Works numbers 8890 and 8891, built in 1918, they are both 8 NHP engines. It is not thought that any other Aveling ploughing engines exist in this country. Attempts to combine the merits of roller and tractor gave rise to the convertible engine. Aveling and Porter built considerable numbers of these engines based on the standard 4 NHP tractor, sets of road wheels and rolls being provided. The roller head was attached to the smokebox with a flanged joint. Both Aveling & Porter and Aveling Barford exported their engines to a great many countries worldwide.

Right Essex County Council owned this neat little Aveling & Porter 4 NHP five-ton tractor. Its duties included hauling road-making materials from pits and railway stations in the county. *Margaret*, a compound engine, works number 7898, was built in 1913. Ickleton, September 8 1962.
Below right Aveling & Porter tractor 9228 of 1921, was supplied new to Hereford County Council. When the engine changed hands it was used on agricultural duties, including threshing. It is a 5 NHP compound two-speed slide-valve engine weighing six and half tons.
Below Aveling & Porter 4 NHP tractor 8288, *Fire Queen*, was built in May 1914. This five-ton two-speed compound engine was one of the first engines to join the Paisley Collection. The engine was resplendent in a smart green livery, when photographed at Chatteris, Cambridgeshire, in July 1963.

A nippy 4 NHP Aveling five-ton tractor, number 12152, was
built in 1928. The engine was taking part in a road run in
Bedfordshire during October 1984 when this picture was taken.

Most of the working life of Aveling & Porter traction engine number 11997, was on timber haulage, apart from the two years from new in 1928 when the engine was owned by Maidstone Rural District Council. This engine was purchased for preservation in 1948, and has been seen at rallies for many years. This photograph was taken at Haddenham in 1971.

Aveling & Porter 4 NHP five-ton compound tractor was built in 1926 and supplied to an engineering company. Photographed on a rather dull late-May day at the 1967 Pirton rally.

Above This very unusual Aveling & Porter vertical-boilered tandem roller made its appearance at the Stourpaine Bushes event, September 15 1984, where it attracted much attention. Photography was difficult with the very large crowds attending the event. The roller is number 12023 and was built in 1928. It is the only known example of the type on the present day rally circuit.

Above left This unusual Aveling & Porter 4 NHP single-cylinder tandem road roller was among the entries for the 1981 'Expo' event.

Below left In 1918 the Aveling & Porter tractor 8809 was supplied new to a Glasgow contractor, later moving south to the county of Cumberland. The engine is a double-crank compound of 4 NHP, weighing six tons. This photograph was taken at the Ely rally held on July 24 1966.

Below A sturdy ten-ton 6 NHP Aveling & Porter road roller, works' number 10372, built in 1922. Several fine examples of this once very-common steam road roller, can be seen at most of the Roxton rallies.

Left An Aveling & Porter eight-ton road roller shortly after being purchased for preservation in 1965. This roller is works' number 10677 built in 1923, and is a 4 NHP engine, supplied new to an owner at Guildford, Surrey.

Right A fine example of an Aveling & Porter single cylinder 4 NHP eight-ton road roller. Number 9155 was built in 1920, spending its early days in the service of a public works contractor in North Lincolnshire. When photographed on July 17 1982, the Aveling was attending the Weeting rally.

Right This superb example of an Aveling & Porter ten-ton compound road roller was built in 1927 as works' number 11793. This engine is typical of the many hundreds of similar engines which once maintained our highways.

Left Aveling & Porter 10437 is a convertible type — note the fixings for extension to the smokebox when used as a road roller. This engine was built in 1925.

Thought to be the only Aveling & Porter showman's engine in existence, works' number 4885, *Samson*, was built in 1901 for the Admiralty who used it at the Chatham dockyard. The engine carried a capstan fitted to the rear axle and was also equipped with two water columns and two pressure gauges in its service days. In its showland days it was owned by Charles Presland but it ended its commercial life, like so many other engines, on agricultural duties.

Top right This typical Aveling & Barford single-cylinder roller is works' number AC 605. It was built in 1937, and is a ten-ton model. For so many years steam road rollers were a familiar sight on the highways and byways, often with a driver who had looked after a particular engine for many years. When moving between jobs, the roller set out under her own steam towing its living van and water cart.

Above right This post-war road roller was built in 1948 by Aveling & Barford Ltd as works' number AH 392. It is a single-cylinder two-speed piston-valve eight-ton roller. When new it was supplied to Newton Le Willows Urban District Council before moving to Redruth in Cornwall. The engine was being prepared at the 1967 Market Bosworth rally.

Right Another post-war Aveling & Barford road roller, AC 757, was built in 1946. This is a six-ton 4 NHP engine.

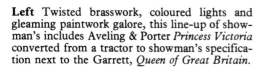

Left Twisted brasswork, coloured lights and gleaming paintwork galore, this line-up of showman's includes Aveling & Porter *Princess Victoria* converted from a tractor to showman's specification next to the Garrett, *Queen of Great Britain*.

Below left The unique Aveling & Porter wagon, pictured here shortly after restoration, *Her Ladyship*, was built in 1921 and supplied new to Gudgegory Council near Sydney, Australia. After changing owners more than once the wagon ended up at a gold sluicing plant, eventually becoming derelict. This wagon was restored in just a few months, making its debut at 'Expo' in late August 1978. It is five-ton compound side-valve wagon and has made appearances at a great many events since 1978, attracting much attention wherever it appears.

This impressive view is of Burrell road locomotive 3057, *Lord Roberts*, an 8 NHP three-speed double-crank compound contractor's engine, somewhat more heavily constructed than normal.

2   Charles Burrell & Sons Ltd

One of the best known engine builders, must surely have been Charles Burrell & Sons Ltd of 'Saint Nicholas Works' Thetford, Norfolk. During the many years that this company was producing steam traction, engines of all types were manufactured.

Many Burrells have survived into preservation, the oldest being a fine traction of 1877, frequently seen during the rally season in various parts of East Anglia. The engine concerned is works number 748, built in 1877, and aptly named *Century*. This engine was probably built as a chain drive, being converted later to gear drive. Like so many other steam engines it stood derelict for a long period, over 30 years in fact, before being carefully restored to the condition in which it can be seen today.

Just two years later, the last surviving examples of the Burrell ploughing engine left the Thetford works. They are numbers 776 and 777, completed in June 1879, and are single cylinder 9 in by 12 in 8 NHP two-speed engines. When purchased by the late Mr Paisley they were in an extremely derelict condition. Over a period of time, work was carried out on number 777 and at the time of the Holywell engine sale in 1980, it had received many new parts, including a boiler barrel and firebox. The winding drum was reconditioned and prepared for use. However, much work remained to be done when both were purchased for restoration at Gisburn. Number 777 was completed first and made its rally debut at 'Expo' in August 1981, carrying the name *The Earl*. The sister

engine is at the present time undergoing the final stages of restoration. Both engines have been sold and will eventually be on show at a museum in Suffolk.

In the field of heavy haulage the Burrells were very popular, especially with showmen, a great many showman's road locomotives were built — quite a considerable number were to special order. As a company, Burrells' engines had a very high standard of finish. Many of these magnificent engines have fortunately survived. The last Burrell showman's built was 4092, *Simplicity*, constructed for Mrs Deakin of Brynmawr, a fine 8 NHP three-speed engine. The engine was actually built by Garretts of Leiston, the Thetford firm having collapsed in the Agricultural and General Engineers combine, causing the last Burrell engine to be built by what had been up to that time a rival firm.

In the case of *Simplicity* the engine's plates still recorded Charles Burrell & Sons Ltd as the builders. Like several other showman's she ended her working days on heavy haulage in the Glasgow area. When she was redundant the engine was intended for preservation but unfortunately this was not to be and the engine was scrapped. Today, given fine weather conditions the Stourpaine Steam Working in Dorest can provide the excellent sight and sounds of large Burrells on heavy haulage, as they tackle the inclines of the site pulling large, heavy timber wagons.

For the areas of Britain with difficult terrain, such as the hilly areas in the West Country, Burrells produced a light traction engine for threshing and other duties. The 6 NHP design was popular and several of the type known as the 'Devonshire' have survived. Being only 6 ft 3 in wide it was an extremely handy engine to get into restricted entrances etc.

Another notable Burrell design was the five-ton steam tractor, which won the 'RAC Gold Medal road vehicle trials'. After winning, the design became known as the 'Burrell Gold Medal Tractor'. In 1983 several of the survivors attended the Weeting rally and made road runs to their original birthplace at Thetford.

Like the other major builders Burrells also produced road rollers, together with a convertible which could be used as either a traction or roller. Burrells designed and built only a limited number of five-ton overtype wagons, some of which were used by showmen, but at least one Burrell wagon is in preservation in this country.

Before leaving the Burrell products mention must be made of three crane engines. Perhaps not the most widely known, but certainly a unique engine, is *Emperor*, number 1876, built in 1895, a 10 NHP single-crank compound road locomotive with two speeds and Burrell's patent spring gear. Although built for showman's use, this engine was works engine at Thetford for many years. Ending up in the Norwich area, it was purchased by Mr Paisley in 1959 and became a regular entrant in the 1960s at agricultural shows and rallies. It was sold in the October 1980 Holywell engine sale. Another well known crane engine is 3829 *His Majesty* a double-crank compound sprung road locomotive going to her first owners at Liskeard, Cornwall in 1920. In October of that year the engine was purchased by Messrs J Hickey and Sons becoming familiar in the London area, until after the Second World War. The final crane engine is 4074 *The Lark* again a double-crank compound, this time with three speeds of 5 NHP. This engine was supplied new in 1927 to a timber merchant at Bury St Edmunds, but it is now resident in East England in the county of Lincolnshire. Fortuantely many Burrells survive, and at least one or two can be found at most rallies.

Right Veteran Burrell agricultural engine number 1244, was built in 1886. Photographed at Westing while undergoing restoration in 1982, this 7 NHP single-cylinder engine was used in Norfolk for many years.

Below left Thought to be the oldest Burrell traction engine in existence, number 748, *Century*, was built at Thetford in 1877. This interesting single-cylinder engine is now preserved not many miles from its birthplace, attending several rallies each year.

Below On a sunny July day in 1963, Burrell double-crank compound light road locomotive, number 4066, awaits the main ring events. This engine was built in April 1927. It is a 5 NHP ten-ton engine originally supplied to a Norwich timber merchant and spent much of its working life for various owners in the county of Norfolk.

Above Burrell agricultural engine and Marshall drum, seen here at the steam working demonstration section of the Weeting rally. For a number of years this rally has presented engines working on timber haulage, threshing, stone crushing and wood sawing with the Fowler ploughing engines operating a steam plough.

Left Burrell 2921, *Violet*, a NHP engine built in 1907 make its way slowly round the show ring. The rally was the 196 Ickleton event held i September.

Above A fine example of a Burrell traction engine, number 3902, *Elizabeth*, was built in 1921, and for its first nine years was employed on timber haulage in the county of Cheshire. Later duties were agricultural, on contract threshing etc. *Elizabeth* is a double-crank compound engine, caught here by the camera at Weeting.

Right A fine example of a Burrell 6 NHP agricultural engine, caught by the camera at the 1979 Haddenham rally. Number 3655, *English Hero*, is a double-crank compound built in 1915. For 20 years this engine stood without protection. It was restored in 1966.

Above One of the veteran pair of 8 NHP Burrell ploughing engines, numbers 776 and 777, built at Thetford in 1879. Purchased for preservation from Wiltshire by Mr Paisley, work was well advanced on a new boiler and restoration of 777, the engine illustrated, when the collection was sold in October 1980. The engine was completed and attended its first event, the 1981 'Expo' held in late August.

Above left Restoration of one of the unique pair of Burrell 1879 ploughing engines was under way at Holywell in 1966. This engine is 777, now fully restored and carrying the name *The Earl*. A new boiler barrel and tube plate had been fitted, the winding drum had been removed for restoration.

Left Burrell ploughing engine 776 was in a sorry state when photographed. This is the second engine of the pair of Burrell ploughing engines. Both of these unique Burrells have now returned to a county adjacent to their birthplace.

Above Burrell traction 4037 is a compound 5 NHP design, built in 1926. The engine is seen here taking part in the grand parade of engines at a rally in 1962, note the absence of large crowds round the ring.

Below Burrell 2789 was built in 1905. Originally a showman's locomotive named *Lord Kitchener*, it was converted for heavy haulage before the First World War. Note the large rear wheels and other embelishments. This engine now carries the name *The President* and is part of the Bressingham collection. *The President* is an 8 NHP engine, weighing 16 tons. Photographed at Ickleton on September 8 1962.

Above Burrells built quite a number of crane engines from 1886 onwards but the one illustrated, number 4074, *The Lark*, was the last built, leaving the Thetford factory in 1927. The engine is a road locomotive fitted with a crane, which has a nominal lift of four tons. Extra strong forecarriage and wheels are fitted. Photographed at 'Expo' in 1978.

Below This interesting Burrell road locomotive was built in December 1919 and exhibited in London, being purchased by an owner in Taunton, Somerset, who used the engine for timber work and threshing, remaining in the same ownership for all the engine's working life. Works' number 3824, *Lord Fisher of Lambeth*, is a 6 NHP double crank compound three-speed engine weighing ten and a half tons—when the engine left the Thetford works it was in showman's colours.

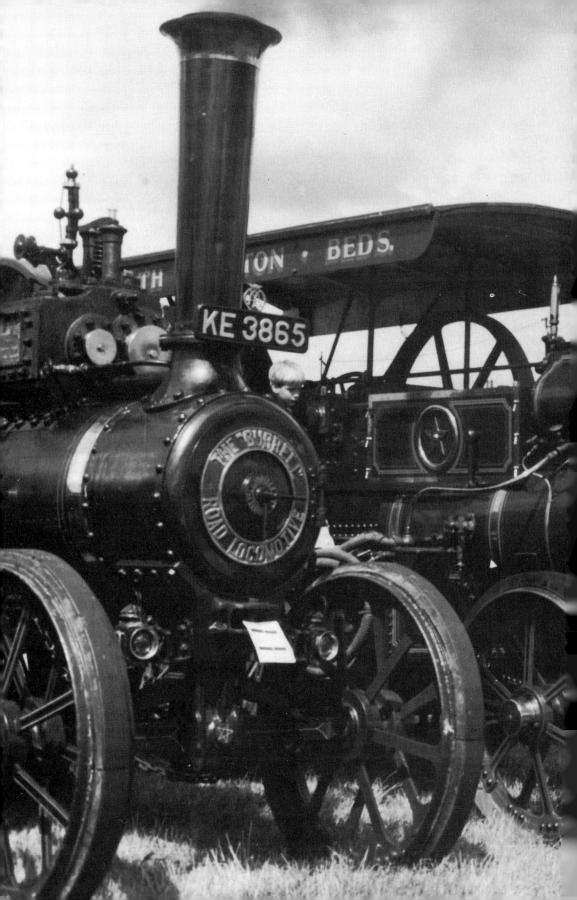

A firm trio of superbly restored engines, photographed in the line-up at the 1981 Weeting rally. The centre engine in the picture is Foster traction number 3710, a 7 NHP unit built in 1908. On the left is Burrell double-crank compound three-speed road locomotive 3593 of 1914, while on the right is Burrell agricultural 2921 of 1907, a compound 7 NHP engine weighing ten tons.

One of the best-known engines in the Eastern Counties is Burrell 6 NHP road locomotive *Duke of Kent*. Supplied new in 1914 to an owner at Chatham in full showman's livery it was fitted in 1915 with a half cab by Burrells. It later saw service hauling timber at a sawmill, eventually ending up with a threshing contractor. It has now been rebuilt to original specifications.

Fortunately ground conditions at the 1984 Stourpaine rally were good, enabling the haulage engines to really get to grips with the hilly terrain. Here Burrell 3057, *Lord Roberts*, and Fowler 17212, *Wolverhampton Wanderer*, near the top of a particular hilly section with a load of timber, having provided a spectacular sight and many enthralling sounds to the large crowd present.

Above A famous Burrell road locomotive, fitted with a jib crane. Note the modified and strengthened front end. Number 3829 was despatched from 'St Nicholas Works' to her first owners in Liskeard Cornwall on March 15 1920. After only a matter of months the engine was purchased by Messrs J. Hickey & Sons Ltd, named *His Majesty* and used for heavy haulage work in the London area and then as a crane engine until 1950. After lying out of use a decision was made to rebuild and restore the 'crane engine' to its former glory. This photograph was taken on September 9 1961 at Ickleton, while still owned by the company as evidenced by the various company embellishments, such as brass ring and star on the chimney.

Right An unusual view of a Burrell single-cylinder eight-ton three-speed engine number 3931 of December 1921. *Mollie* was photographed at the 1982 Roxton event shortly after a repaint. The engine was steamed up from Kent on more than one occasion, attending both 'Expo', and Roxton rally on its return south.

Left *Peter Pan* a Burrell five-ton showman's tractor. This 4 NHP engine was exhibited at the Smithfield show when new in 1912. Converted to a showman's in 1928 and travelling thereafter for a showland owner based at Chichester Sussex, this double-crank compound engine is Burrell works' number 3433.

Left Burrell three-speed 'Gold Medal' tractor number 3618, *The Scout*, was supplied new to an owner in Lancashire. This neat engine was built in 1914, and is a 4 NHP double-crank compound weighing five tons.

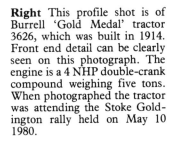

Right This profile shot is of Burrell 'Gold Medal' tractor 3626, which was built in 1914. Front end detail can be clearly seen on this photograph. The engine is a 4 NHP double-crank compound weighing five tons. When photographed the tractor was attending the Stoke Goldington rally held on May 10 1980.

Left For many years this Burrell showman's tractor toured the fairgrounds on the Isle of Wight. Burrell 3868, *Island Prince*, was built at Thetford in 1920. It is a 4 NHP double-crank compound five-ton tractor. Before the engine moved to the Isle of Wight it was owned by a showman at Southampton and in those days carried the name *The Russell Baby*. Photographed at Great Wymondley on June 6 1965.

Above Three showman's locomotives, caught in a quiet moment at the 1984 Haddenham rally. Nearest the camera is Burrell 3933, *Princess Mary*, a 7 NHP showman's supplied new to John Anderton of Exeter but later moving into the county of Surrey. Next to the Burrell is Garrett 4 NHP showman's tractor 33987, *Woofy*, built in 1920. At the end of the line is the magnificently restored Burrell 3979, *Earl Haig*. This last engine, built in May 1924, is owned locally and is well known at this, and many other rallies throughout East Anglia.

Below A showman's engine not seen at many events, Burrell 6 NHP number 3878, *Island Chief*, built in February 1921. This Burrell is a 12-ton double-crank compound three-speed engine. Photographed at the 1962 Raynham day, Norfolk.

Charles Burrell & Sons Ltd

Burrell showman's 3884, *Gladiator*, constructed at Thetford in 1921, photographed here still in the process of being restored to its former glory. This 8 NHP engine is a double-crank compound three-speed engine weighing 20 tons. For many years the Burrell was used to drive a scenic railway.

A Burrell showman's locomotive once familiar on the northern fairs circuit was number 3610, *William V*. This 8 NHP three-speed engine was a regular entrant to rallies in the Bedfordshire-Hertfordshire area during the 1960s. Note the bracket for the exciter generator, which was not fitted when photographed at Houghton Conquest, September 22 1968.

Ready for home. Burrell 3526, *Lightning 11*, has been loaded and will shortly be ready for off when the trailer wheels are added. This engine was built in 1913, a compound three-speed 7 NHP engine, originally travelling to the fairs in Cumberland and the surrounding districts.

Gladiator is a Burrell 8 NHP scenic showman's double-crank compound three-speed engine built in 1921 as works' number 3884. Supplied new to a London-based showman.

In 1907 this fine Burrell showman's was supplied new to Henry Thurston, the famous Eastern Counties' showman, and was known as *Lord Nelson*. When the engine was travelling in the south it became *Princess Royal*. The engine is a 7 NHP compound built at Thetford in 1907. Seen here at the Herts Steam Preservation Society rally held at Great Wymondley.

Black Prince, the Burrell showman's from the Bressingham Collection. Number 2701 of 1904 is an 8 NHP showman's road locomotive weighing 18 tons. After a period of lying derelict this engine was restored in 1963. Among its last duties was the demolition of 'blitzed' buildings in London.

Above Burrell 3334, *The Bailie*, supplied new to Greens of Glasgow in 1911, ended its working days travelling the Lancashire fairs until 1945. After a period of storage it was acquired for preservation. Photographed at Rempstone (Nottinghamshire) July 2 1967.

Below *Quo Vadis*, a Burrell scenic showman's locomotive of 8 NHP, built in 1922, makers' number 3938. Supplied to W. Wilsons a London-based showman for generating power for his Rodeo switchback, the engine was retrieved for preservation in an almost hopeless condition and restored to its former glory.

Above Sunday morning and the showman's engines receive a polish, ready for the public. Here 8 NHP Burrell 3443, *Lord Nelson*, is prepared. This engine was built in 1913 and supplied to the west of England amusement caterer Anderton & Rowland.

Above right A magnificent example of an 8 NHP Burrell showman's road locomotive, *His Lordship* was delivered in 1913 to Greens of Glasgow to power their 'Dragon Scenic Railway'. It later went south to Preston and finally to Silcocks of Manchester.

Below right A fine example of a Burrell showman's engine. Number 3526, *Lightning 11*, built in 1913. This engine was one of the last to be used on heavy haulage. The overhead power lines in the background, indicate that the photograph was taken at 'Expo'. August 26 1978.

Below Surrounded by fairground rides and organs at the 1967 Castle Howard Steam Fair and Rally, Burrell 8 NHP scenic engine 3909, *Winston Churchill*, was built at the famous Thetford works in 1922. This engine travelled originally for Messrs Holland in the Midlands.

Above When *Earl Haig* left the Burrell works in 1924, the engine was supplied to a showman who travelled the West County fairs. Like so many other showman's engines *Earl Haig*'s final days were spent on threshing etc. When recovered for restoration the engine was in a very derelict condition but has been restored over the years to the magnificent condition that the engine is in now.

Below Burrell 8 NHP showman's scenic 3887, *The Prince of Wales*, was supplied new to a west of England amusement caterer based at Devizes, Wiltshire. The engine provided power for a famous set of 'Whales' scenic rides'. The photograph was taken at the 1984 Stourpaine Bushes event. For many years the engine attended the October fair held at Salisbury, not very far from Stourpaine.

Above The writer of the introduction to this book, photographed with his Burrell showman's locomotive 3555, *The Busy Bee*, built at Thetford in 1914. This engine is a double-crank compound three-speed locomotive of 5 NHP. Supplied new in March 1914 to Taylor Brothers, Workington, Cumberland, and travelling mainly in south-west Scotland, Cumberland and parts of Yorkshire, she ended her working days in Northumberland on threshing and agricultural work. Dr Middlemiss purchased the engine while in residence in Northumberland and when changing practice, the engine was steamed some 250 miles south.

Below Burrell scenic 10 NHP showman's locomotive number 4000, *Ex - Mayor*, was supplied new to G T Tuby & Sons, Doncaster. The engine has been overhauled and is now in superb condition, returning to the rally fields in 1984. In October it was photographed at the end of the season steam-up.

This side view of the cab of Foden 'D3' three-speed tractor 13222, enables details to be clearly seen. *Cheshire Maid* is a much-travelled Foden attending many engine events in different parts of the country.

3   Fodens and Yorkshires

One of the most successful overtype steam wagon builders was Messrs Fodens of 'Elworth Works', Sandbach, Cheshire. This company, with its principles of economy and efficiency built up a strong following with many steam operators. The distinctive outline of the Foden wagon and tractor with its locomotive-type boiler makes them easily recognisable.

It is for wagons that the company is best known particularly the famous five tonner range with the overtype engine mounted on top of the locomotive-type boiler. As in conventional traction engine practice, the drive to the rear axle was by heavy chains. In the early 1920s the six-ton wagon was introduced, incorporating many developments designed to improve operation on the road — three speeds, Ackerman-type steering, an improved cab, and the wagon also required less oiling on the road. The official speed limit of 12 mph restricted these wagons, they were certainly capable of much more.

The company also built traction engines and road locomotives, the first traction being built in 1880. From the start the company adopted principles of quality and durability. Shortly after the first traction was introduced all engines including tractions were designed with springs. Production of traction engines lasted until 1920 when the company decided to concentrate on road vehicles.

Over the years the company exported a great many of their engines and wagons to many countries, particularly to South America and South Africa. They also sent considerable numbers to Australia where they were widely used on timber haulage. One Foden was recently returned from Australia and restored in Britain, before being sold and going on to a German steam enthusiast.

When the First World War broke out the government War Department commandered the whole output, the only exception in this country being listed priority users. With the works producing flat out the Fodens were loaded on to rail at Sandbach and shipped straight over to the Army in France. After hostilities ceased, the home market was flooded with these ex-War Department Foden wagons, causing problems to Fodens and other wagon manufacturers, with many wagons of several makes being available at low prices.

Some were converted to steam buses but they were never very common. Nevertheless almost all the steam wagon builders did at least look into bus development. One Foden steam bus was used to transport the famous Foden Motor works' brass band to their various engagements. The company did produce undertype designs from 1926 but none have survived into preservation, the designs being killed off by the 1930–34 Road Traffic Acts.

Over 43 Foden wagons exist today with various bodies including flat open-type and three-way tipping bodies, while others have tank bodies fitted. The Foden tractor has 25 examples of different types in preservation. A sole multi-axle wagon exists, while in Lincolnshire an interesting 'Sun' tractor has been restored.

Several traction engines and road locomotives exist, together with one Foden showman's road locomotive, works number 2104 *Prospector*. A replica of the brass band works' bus can occasionally be seen at events in East Anglia.

The early wagons had an outside frame chassis. The cab offered very little protection and, of course, they were fitted with solid tyres all round. Over the years the cab was redesigned to give much greater protection,

the final wagons being fitted with a much improved cab and also pneumatic tyres.

Foden tractors were popular engines. Those designed for use with timber had a short body to enable them to manoeuvre in confined spaces and a winch was often fitted for winching out trees. Some were fitted with rubber tyres for use on the highway.

The restrictions on steam wagons introduced in the 1930s forced many steam haulage contractors to dispose of their Foden wagons, changing over to internal combustion vehicles.

Right An early five-ton Foden wagon, works number 1742 was built in December 1908. The wagon was lettered in the name of the Tiverton brewers and carries a dray-type body. Several traction engines and wagons were at the Tyseley Open Day on September 29 1968.

Below right Foden three-way hydraulic tipper number 13716 was built in 1930. This wagon was supplied new to Northamptonshire County Council. For many years the wagon lay derelict before being rescued for preservation. After restoration the wagon has travelled a great many miles under its own power to and from various events.

Below This Foden five-ton wagon was awaiting restoration when photographed at the September 1961 Ickleton rally. Works' number 8304, it was built in 1918. At one time this wagon was owned by the War Department and later it was owned by Croydon Council, being the last Foden used by them.

Above One of the last steam wagons built by Fodens was number 13708, built in 1930. This six-ton wagon is now named *Boadicea*, and is included in the Bressingham Collection. In its later years it was used as a tar sprayer. Photographed at Thetford on July 30 1967.

Below Foden six-ton wagon number 10788, built in 1922. This wagon attended the first 'Expo Steam' held at Peterborough in 1973.

Above A much-travelled five-ton Foden wagon, makers' number 3510 was built in 1913. In 1927 the engine was purchased by Henry Ford for his museum at Dearborn, Michigan, USA. Much of the paintwork on this Foden is original and since it returned to Great Britain it has been seen at many events.

Below This superbly restored five-ton Foden wagon was seen at the 'Expo Steam' event held at the East of England showground, Peterborough. Works' number 5078 was built in 1914.

Above Several steam wagons built by both Sentinels and Fodens attended the Harewood House rally, held in the grounds on September 19 and 20 1964. Among the entrants was Foden 12364 of 1926. In its last working days this Foden was employed as a tar sprayer.

Below This neat six-ton Foden tractor was taking part in the 1967 Market Bosworth rally. Built in November 1928 this 4 NHP engine was employed for some years on coal haulage in the London area.

Above Foden 'D' type tractor number 14078, *Mighty Atom*, was built in 1932. This compound, 4 NHP eight-ton unit is one of the last Foden steam tractors built to survive in preservation. It was supplied new to a timber merchant, then worked with a set of fairground roundabouts for a number of years.

Below Foden 13784, a 6 NHP eight-ton tractor built in 1931, was fitted with pneumatic tyres on the front and solid tyres at the rear, when photographed at the Tyseley Open Day in September 1968.

A much-travelled Foden. Number 4086 of 1913, *Her Grace*, a six-ton Colonial wagon built in November 1913. The wagon was shipped out to Australia but after many years it was purchased derelict with many parts missing and returned to England in a crate. Superbly restored and finished with yellow paintwork, it attended its first rally at 'Expo' in August 1980. This Foden has a boiler modified for wood burning and is thought to be the last example of its kind. The wagon was sold in August 1984 and is now in Germany.

Above Foden tractor 13008, *Wanderin Tam*, built in July 1928, seen here at Castle Howard in August 1980. Note the twin rear wheels and twisted brasswork on this rare six-wheeled tractor. For many years this Foden was owned by Trinidad Liverpool and used on tar spraying duties.

Above right This superb Foden is a familiar sight at rallies and vintage vehicle events. Foden works' number 13222 is a Class 'D3' three-speed slide-valve compound tractor of 4 NHP, weighing eight tons and built in 1928. This tractor, which is fitted with pneumatic tyres all round, travels to events under her own steam.

Below right Another unusual and unique tractor is the Foden 'Sun' two-cylinder tractor, Number 13730, built in 1931. The vehicle's motion can be seen clearly in this photograph of this extremely interesting engine, which is not a frequent sight on the rally fields.

Below *Castria*, a Foden 4 NHP tractor, 13218, was built in 1928. Purchased for preservation from a sawmill in the mid 1950s this smart tractor was photographed at the 'Great Yorkshire Steam Fair' held at Castle Howard on August 9 and 10 1980.

Above Bright July sunshine reflects on the paintwork of Foden tractor 13068 built in September 1928

Below Foden 11340 built in December 1923, a replica of the vehicle that was used to transport the famous 'Foden Works Band' in the years 1916–23. This vehicle was providing rides at a steam rally held at Bressingham on August 28 1967.

Above One of the few remaining Foden road locomotives, number 1294, was built in 1909 and is a compound 7 NHP engine of which little history is known. When purchased this engine was in a very derelict state, new motion, crankshaft and fittings being required to reach the fine condition that this vehicle now demonstrates.

Below Foden showman's road locomotive, *Prospector*, was built in 1910, as works' number 2104. This compound 6 NHP engine is rarely seen in Eastern England. However, it has on one occasion, attended the 'Expo Steam' event held at Peterborough, where this photograph was taken.

Yorkshire Patent Steam Wagon Company

The city of Leeds, the location of several engine building companies, included one of the largest, John Fowler & Co (Leeds) Ltd. Leeds was the birthplace for many hundreds of traction engines and railway locomotives. With the large export markets that several of the companies built up over the years, Leeds products reached a great many distant and isolated parts of the world.

One of the manufacturers, located in Pepper Road, Leeds, was the Yorkshire Patent Steam Wagon Co. Not many of this company's products have survived, but what has is extremely interesting. The oldest and smallest of the Yorkshire wagons to survive is a two-ton, 'W' Class unit, built in 1905. Two other wagons, built in 1914 and 1917 respectively, represent the older designs. These are Class 'WE' and 'WA'.

The latter unit, number 940, has an interesting history although it never strayed far from Leeds, and we are fortunate that this vehicle is with us today. This three-ton wagon was supplied new in 1917 to a Leeds engineering company but eventually changed hands and was used for scrap metal haulage. In 1931 the wagon was laid aside, and slowly became covered by scrap metal. In 1961 it was cut into pieces. When discovered it was regarded as past rebuilding. Many, many hours of careful and painstaking work resurrecting the wagon have resulted in the fine example which we see today.

A six-ton 'WE' Class wagon, works number 2108 built in 1927, and a tractor unit from an articulated wagon, works number 2118, again built in 1927, are the last Yorkshires built to have survived to the present day, as far as known. One other incomplete Yorkshire wagon has survived, and hopefully, this wagon too, will be seen at rallies in the future.

Yorkshire 2008 of 1927, *Pendle Laddie*. This six-ton tractor was originally built as an articulated unit for Leeds Electricity Department and only later rebuilt as a ballast tractor.

Photographed in close up is the transversely-mounted boiler of Yorkshire two-ton wagon number 117 built in 1905. Details of the boiler and mountings can be clearly seen in this photograph, taken in 1973.

The earliest known surviving Yorkshire, is this two-ton 'W' Class wagon, which was built at Leeds in 1905. This wagon is typical of the period in that no protection is provided for the driver.

Above A remarkable wagon which we are fortunate to have in preservation, this unusual Yorkshire three-ton wagon was built at Leeds in 1917 being used by various owners in the area with the last using the wagon for scrap metal haulage. Eventually the wagon was laid aside, gradually becoming buried by scrap metal until the early 1960s when it was cut into pieces. Many people considered the wagon past restoration. However, it has been superbly restored and now attracts considerable attention wherever it goes.

Below Yorkshire overtype wagon *Yorkshire Lad* built in 1927, works number 2108, model WG. This wagon was rebuilt from a very derelict state, and is now one of the six Yorkshire engines known to have survived. Photographed at the 'Expo' event held annually at the East of England Showfield, Peterborough.

4   John Fowler and Co (Leeds) Ltd

This Leeds-based firm is one of the most famous steam builders, its products being supplied to many parts of the world. Fortunately a great many 'Fowlers' are in preservation in both Great Britain and several other countries. As a company Fowlers were never short of ideas, and what's more, were never afraid to try them out. Originally in business for ploughing tackle, the steam plough works started production in 1861. Within a matter of four or five years the works progressed on to producing traction and portable engines.

Several very early ploughing engines have been carefully restored. The three oldest survivors were built at Leeds in 1869-70-73, all are 20-ton engines, and another seven date from 1874-84. Quite a number of years are unrepresented before the 'B4' ploughing engine built in 1910, followed by the sole surviving 'K5' which is a single-cylinder 8 NHP ploughing engine, named *Old Guard*, works number 12366. This particular engine was found in a scrapyard in Hertfordshire in 1960 in an extremely derelict condition. After 1913 the numbers of Fowler ploughing engines to survive increases considerably and several classes exist, the most common being Class 'AA' and 'BB'.

Many ploughing engines were exported to Europe and many much further afield. In recent years some of these engines have returned to this country including several Class 'Z7' engines after years in the scorching African sun. These were the standard ploughing engine, originally exported to the Sena Sugar Estates Ltd, Mozambique. The 'Z7's are large compound engines weighing 25 tons, mostly fitted with full length canopies and 6 ft chimneys. Large rear wheels of 7 ft diameter, and 3 ft wide were fitted with narrow strakes.

Even in this modern age Fowler ploughing engines still see occasional commercial use, at least one pair still carry out dredging work on ponds, lakes and reservoirs. Such is the interest in steam power operations, even in this modern age of speed and electronics, that the power of steam attracts onlookers and television programme producers alike.

Fowlers are also well known for their showman's road locomotives, 73 being adapted for use by travelling showmen, the bulk being within the period 1900-14. The first Fowler showman's was built in 1886. The last showman's built also survives. Works number 20223 *Supreme* was built in 1934 for a Welsh amusement caterer. This engine was built to special order with left hand steering and chrome fittings instead of brass. *Supreme* was used on fairgrounds until World War 2 and then used for heavy haulage in the Glasgow area.

At the end of World War 1 considerable numbers of Fowler road locomotives became available, no longer required for War Department service. Many were converted to showman's road locomotives, while others found their way into heavy haulage.

Showmen achieved surprising haulage feats — their wagon trains, which could be up to ten vehicles, often had a combined weight of nearly 50 tons. Under good conditions and given reasonable roads, speeds of 15-20 mph could be reached. Average speeds of 10 mph over short hauls were not uncommon. The truly unforgetable sight and sounds of a showman's road locomotive getting to grips with its load, especially on a slightly adverse gradient, are regretably now memories of the past.

Heavy haulage was a duty on which the large Fowler road locomotive really excelled. Some firms such as Norman Box, who were specialists in heavy haulage, retained a fleet of large Fowlers' road engines. These engines would be called upon to move jobs such as heavy boilers, timber, engineering products, bricks etc. Fortunately several of these heavy road locomotives are now preserved in excellent condition and it is especially pleasing to see them on occasion performing the tasks for which they were designed. Weather conditions permitting, the annual Stourpaine event sees the heavy haulage engines at the head of considerable timber loads on the steep inclines of this Dorset site. Another very popular Fowler design was the nippy little 'Tiger' five-ton tractor. These engines could be operated by one man and were mainly used for light haulage duties.

Fowlers produced single and compound general purpose traction engines in considerable numbers in the period up to World War 1, after which the steam scene was mainly road rollers and ploughing engines.

Considerable numbers of road rollers were built by the company. The earliest survivor was constructed at Leeds in 1896, it is a 12-ton single-cylinder design of Class 'D1'. Production continued through to the 12 and 14-ton compound rollers of 1937–38.

One unique Fowler six-ton wagon survives. Works number 19708, supplied new in 1931 to the City of Leeds cleansing department as a gully emptier, was scrapped in 1941. After spending 20 years buried by scrap metal, it was cut into pieces and it was in this form that it was purchased for preservation in 1968, with many parts missing. It can now be seen on the rally field in superb condition, an extremely rare, interesting and historic vehicle. Nearly 500 engines of various types built by Fowlers are in preservation in the British Isles. More still survive in other parts of the world, some in preservation, but other rollers abroad are still performing the duties for which they were built.

Veteran 20-ton Fowler single-cylinder ploughing engine number 2013, built in May 1873. Only two ploughing engines older than *Noreen* are known to exist, being built in 1869 and 1870 respectively. This engine attended the 1978 'Expo' event at Peterborough.

Above Already over one hundred years old when photographed, this Class 'A' Fowler 13-ton single-cylinder ploughing engine of 12 NHP, left the Leeds works way back in 1875. Several Fowler ploughing engines built in the 1870s have survived, the oldest being built in 1869.

Below The second oldest surviving Fowler ploughing engine seen working at the 1984 Stourpaine Bushes 'Great Working of Steam Engines'. The engine had recently undergone major repairs and is now in superb condition. Fowler 1368, *Margaret*, was built in 1870. It is a single-cylinder 12 NHP 20-ton engine.

Above Another veteran Fowler ploughing engine is works' number 2479, built at Leeds in December 1874. This engine is a 12 NHP 13-ton Class 'AA' single-cylinder engine. When photographed the engine was at the Rempstone rally.

Below The sole surviving Class 'K5' Fowler ploughing engine, works' number 12366, was built in 1910 and supplied new to a ploughing contractor in Sussex. The engine is an 8 NHP single-cylinder fitted with a two-speed ploughing gear, the engine carries the name *Old Guard*. When caught by the camera at Wicken near Ely it was attending a late-season ploughing and vintage machinery event.

Above A pair of Fowler Class 'BB' 20-ton ploughing engines photographed at Carrington Park rally held near Boston, Lincolnshire. The engines are works' numbers 15441/15442 built in March 1920, and carrying the names *Tiger* and *Lion* respectively.

Below Another view, showing the other side of the rare Fowler 'K5' ploughing engine, *Old Guard*. Note the front wheels which are well forward of the boiler. The ploughing gear fitted is a two-speed type.

A front-end shot of Fowler ploughing engine 15441 a Class 'BB' 16 NHP 20-ton engine, named *Tiger*. Its sister engine is 15442, *Lion*, and the pair frequently attend rallies. Note the width of the front wheels fitted on these typical ploughing engines.

Above Fowler ploughing engine number 15139 is prepared for the day's events at the 1965 Great Wymondley rally. Another of the many Class 'BB's which are around today, this particular engine was built at Leeds in 1918 and was in use until the late 1950s.

Below Class 'BB' Fowler ploughing engine of 16 NHP number 15207, weighing 20 tons and looking most impressive at a Roxton Park rally. The winding drum and gear can be very clearly seen in this photograph. Quite a considerable number of Class 'BB' engines are in preservation.

Fowler ploughing engine 13881 of 1917, gently raises steam at Rempstone. This engine is a Class 'AA6' compound weighing 20 tons.

In the early rally days many engines were steamed to and from events, some journeys involving the engine and crew being on the road for a couple of days, or more. In recent years the majority of engines are transported on low loaders, travelling on vehicles such as the one in the photograph, from which a Fowler ploughing engine is being unloaded.

One of the massive Fowler Class 'Z7' ploughing engines attending the 1982 'Expo'. Here number 15670, built in 1922, moves slowly towards the demonstration area. This engine is one of a batch returned in 1977 from the Sena Sugar Estates in Mozambique. Note the very tall chimney and wide rear wheels.

Roxton Park, Bedfordshire, among the ploughing
engines preparing to give a demonstration was
this Class 'BB' 16 NHP Fowler, built at Leeds in
1917, number 14383. Note the wire cable has
already been partly unwound from the drum
on *Prince*.

Excelsior and *Captain Scott*, two powerful road locomotives stand together under threatening skies at the 1980 Rempstone rally. The engine nearest the camera is Fowler 15323, *Excelsior*, of 1918 and Class 'A1' compound 7 NHP, weighing 14 tons. The McLaren is works' number 1421 of 1913, a compound three-speed 8 NHP engine, weighing 11¾ tons.

Above left The massive proportions of Fowler 'K7' compound ploughing engine 17757. This 14-ton, 12 NHP engine is one of a pair owned by a museum. Here the engine is raising steam on Saturday July 12 1980 at the Rempstone rally.

Left Fowler compound road locomotive, Class 'D2' of 5 NHP built in 1910 and named *Princess*. Photographed at Stourpaine Bushes on September 23 1972.

Above Fowler 'B5' Class road locomotive 8903 of 1900, seen here at Manor Farm, Holywell shortly after restoration in June 1966. This engine was supplied new to Market Lavington Brick Works—an impressive engine capable of handling very heavy loads.

Right The impressive lines of the Fowler 'B5' road locomotive are seen to advantage in this low angle shot of 8903 built in 1900. *Lord Roberts* is a fast and powerful three-speed double-crank compound engine of 8 NHP.

Above Fowler 14910 was built in 1917 for the Ministry of Munitions, as a haulage and winding engine, the winding drum was mounted under the boiler and removed in later years. Photographed at the 1984 Weeting rally.

Below Fowler haulage and winding engine, works' number 14950, was built for the War Department for the Russian War, but not delivered. It is a Class 'TE' two-speed 8 NHP double-crank compound. On the road these engines were not particularly fast, but very powerful. In 1964 14950 was restored in original colours. Here it is seen at Hinchingbrooke Park, Huntingdon, while attending the Hunts County Show on June 17 1967. Note the winding drum fitted.

Above A front-end study of two interesting engines. Nearest the camera is Fowler 14910, *Rising Star*, built in 1917, a 7 NHP class 'TE2' compound weighing 12 tons. In the background is Ransomes, Sims and Jefferies Ltd number 30004, built in 1919, a 7 NHP single-cylinder agricultural engine.

Below Fowler *Super Lion* crane engine. This sturdy 8 NHP Class 'B6' road locomotive was built in 1929. It is a double-crank compound weighing 21 tons. For many years this engine delivered Lancashire boilers all over the country. Among its last duties was that of yard engine. Since being purchased for preservation this majestic Fowler has travelled extensively to events in many parts of the country and also to Holland. Photographed June 25 1983 at Sudbury.

Above Fowler 'R' Class traction engine, *Jezebel*, was caught by the camera at a quiet moment, during the first 'Raynham Day' held on September 15 and 16 1962. The grounds of Raynham Hall, Norfolk, provided an excellent venue for the two large rallies held there.

Below Standing ticking over at Roxton Park rally, number 11357 was built in 1908. This engine once worked for a familiar contractor, Messrs Kitcheners of Potton. The Fowler is a Class 'A4' single-cylinder 6 NHP engine.

Above Fowler 'R' Class traction engine, number 11594, built in July 1908 and named *Margaret*. This fine 7 NHP single, has been coupled up ready for a demonstration at the second 'Raynham Day', held on September 14 and 15 1963.

Below Double-crank compound 7 NHP Fowler agricultural, number 10373 is an 'R' Class engine built at Leeds in 1905. The engine was supplied new to Stetchworth Hall Estates, spending over 40 years working for the same employer.

Above Two interesting Fowlers. On the left is *Tiger*, a five-ton tractor, 15632, built in 1923. The other engine is works' number 17287, a class 'DNB' compound traction built in 1927 which was originally built as a road roller. **Below** The unique Fowler six-ton wagon, number 19708, was built in 1931 and supplied new to the City of Leeds as a gully emptier. In 1941 the wagon was withdrawn, ending up in a scrapyard and covered with scrap. In 1961 the wagon was cut up and it was purchased several years later as a pile of parts — from which the interesting wagon was rebuilt.

Above When this neat little Fowler tractor, *Tiger*, was purchased by its present owner for restoration, it was little more than a pile of parts. A major rebuild was carried out, including a new boiler and firebox. Work was also undertaken on the bearings and gears resulting in the fine restoration job, which we see today. This tractor, makers' number 15632, was built in 1923 and supplied new to a Scottish owner for timber work. It was later converted to a roller in a Scottish locomotive works, eventually being re-converted to a tractor once again. This speedy engine attends a great many rallies under her own steam.

Right Fowler 15629, *The Tiger*, built in 1920 is a Class 'T3' compound five-ton tractor. Most of this engine's working life was spent in Scotland and Cornwall hauling timber. Photographed at Weeting several years ago.

Left Fowler 'Tiger' showman's tractor 14798, *Firefly*, seen here at Pirton on June 3 1968. This engine's first owner was Dulverton Council. After changing hands it was converted to showman's condition and was used to drive a juvenile roundabout.

Left Quite a number of road rollers were still in commercial use in the 1960s. This 1924 vintage Fowler 'D5', number 16100, ten-tons single-cylinder roller, was still in use near Barmouth on June 17 1963. Several items had been removed, including the Fowler name over the front roll, while the canopy too had seen better days. **Below left** A happy ending for Fowler 16100. Just six years after the previous photograph was taken, the roller was on display at the Banbury rally (June 1969). This roller is thought to have spent most of its working life in mid Wales.

Above right Close up detail on ten-ton Fowler road roller 16134, a compound engine built in 1924 of Class 'D9'. The roller was last used commercially while in the ownership of a company operating from Reading.

Right A Fowler ten-ton single-cylinder roller of Class 'DNA' built in 1929, works' number 17560. This roller was purchased derelict from Kitcheners of Potton, a company who at one time operated many engines.

Above Single-cylinder ten-ton Fowler roller number 15942 of class 'DH1' was built in 1923. The roller is seen here at a steam rally held in 1973.

Below In 1924 number 16236 left the Fowler works as a road roller of class 'DN1'. In later years it was rebuilt as a showman's road locomotive and the engine now carries the name *Lady Mary*. Photographed on a rather gloomy day at the event held in the grounds of Castle Howard on August 9 and 10 1980.

Above A superb example of the Fowler showman's road locomotive, number 15117, *Headway*, was built at Leeds in October 1920. This engine is an 8NHP of Class 'R3', a compound type weighing 13 tons.

Below On a dull and wet day at Castle Howard in August 1980, Fowler 'B6' showman's road locomotive 19783, *King Carnival II*, provides power for the organ, seen in the background with its admiring audience. Built in 1932 this engine was delivered to Frank McConville at Newcastle Town Moor in July of that year. Subsequent years saw the engine on heavy haulage both in London and working for a contractor in Wolverhampton. After many hours of work this fine engine has been restored to its former glory.

Above *Sir John Fowler:* this compound showman's road locomotive of class 'R1' was built at Leeds in 1905, works' number 9393. This particular engine is not often seen in East Anglia, however, on the late summer bank holiday it attended the 'Expo 78' in company with many other fine showman's engines.

Below This Fowler engine spent its commercial life in agriculture, but has been completely rebuilt and converted to full showman's specification. Photographed here providing a truly magnificent sight at the 1984 Stourpaine event, the engine is now named *Monarch of the Road.* It is a 7 NHP built in 1917.

Above Fowler 20223, *Supreme,* the last showman's engine built by Fowlers, leaving the Leeds works in 1934 for Welsh amusement caterers Messrs A. Deakin & Sons, Brecon. The engine was built to special order with lefthand steering and chrome fittings instead of brass and ended her working days on heavy haulage in the Glasgow area. This engine is a 'B6' compound 10 NHP locomotive, weighing 11 tons.

Below A superb Fowler 'B' Class showman's road locomotive 14425, *Carry On,* built in 1916. This engine is a 10 NHP double-crank compound with three-speed gearing. *Carry On* was one of the last showman's in everyday use, retiring in 1959 after travelling for Cadonas of Glasgow and McGiverns in Eire.

The impressive front end of the sole remaining Sentinel 'S8' steam wagon, this particular 'S8' was the works' demonstration vehicle. When purchased for preservation all that remained was a derelict chassis. At one time this wagon was used for hauling steel plate from South Wales to Fords at Dagenham. Number 9105 was built in 1934. This photograph was taken at Roxton Park in September 1984.

5 Sentinel (Shrewsbury) Ltd

Nearly one hundred Sentinel wagons and tractors are preserved in Great Britain. The oldest survivor dates back to 1914, while the youngest was built in 1937. At most steam engine events one can be fairly certain of seeing examples of what is one of the best known steam wagons.

The Sentinel wagons fall into four types namely the 'Standard', 'Super' and later 'DG' and 'S' Series. In addition there are at least 14 tractors. The tractors include timber haulage and winching units as well as haulage units for use with trailers.

The earliest type of Sentinel was the 'standard' model, produced over the period 1906–23. This undertype wagon proved to be most successful in its day. The unit is chain-driven to the rear axle. The cab is rather exposed with open front and sides. Several standard wagons were in regular use at Brown Bayley Steel, Sheffield until withdrawn from use in 1970. Some of the units spent the whole of their working lives at the works. Amongst their duties was moving hot metal to the rolling mills but after many years of extremely hard work several of the Sentinels used at the works are now preserved, and can be seen from time to time on the rally field.

The prototype 'super type' Sentinel made its appearance in 1923 but followed the basic principles of its predecessors in having a vertical boiler mounted at the front with an undermounted engine and a separate chain drives running to each rear wheel. On most units a more weatherproof cab was fitted, thus improving the operating conditions considerably. Solid rubber tyres however, were still the order of the day.

In the late 1920s the 'DG' series replaced the 'Supers' and within this period there was rapid development. The 'DG' series was in production for approximately seven years. The 'DGs' were powerful units working at a steam pressure of 275 psi. They were chain driven, many starting out on solid rubber tyres. The 'DG' series were either four or six-wheel designs. However, in September 1930 the first eight-wheeled design was introduced. Eight new units were produced and a further eight were converted from 'DG6' models. The design was considerably before its time, both in conception and the law. Unfortunately no 'DG8' have survived into preservation, although several 'DG4' and 'DG6' examples are regular rally engines.

The final design was the 'S', or shaft-driven four-cylinder model. The earliest survivor was built in 1933. Sentinel 'S' Series are magnificent vehicles, which can even today, fully loaded, compare favourably with a truck of similar capacity. Numerous improvements were made in the design, 'S' series vehicles were capable of 60 mph, with economical fuel consumption. The appearance was changed with flat-fronted modern cabs. Cab doors incorporating sliding windows, full windscreens and electric wipers were fitted. The lighting was all electric and with pneumatic tyres it all added up to a much improved unit. Examples of the four and six-wheel design have been preserved, as has a sole 'S8'. This eight-wheeler is a regular entrant at rallies held in many parts of Britain.

Tractors built by Sentinel include units designed for use with timber and some are very powerful units. Sentinel 8756, *Brutus*, built in 1933 and now part of the famous 'Bressingham Collection' is an 11-ton tractor fitted with two 120 NHP engines, while others have steam operated winches capable of hauling very heavy items. The tractor and trailer combination, as used for road haulage, can also be seen today at rallies. The Sentinel wagons were popular with fleet owners, some having upwards of 40-50 vehicles.

Regulations against steam vehicles came into force in 1930, in effect making the steam vehicle uneconomic against the rapidly increasing numbers of internal combustion engines. The regulations caused considerable numbers of Sentinels to be consigned to the scrap dealers, while others were converted to tractors suitable for use with trailers, some of the tractors surviving until the 1950s.

Many Sentinels were available in the 1930s, at very low prices, often in an excellent condition and with many more years of life in them. Amongst the most unusual Sentinels to survive is 5644, *The Elephant*, an interesting 'Super' tractor, which was, for many years, responsible for shunting railway wagons on Teignmouth Docks in South Devon. A replica steam omnibus, three-way tipping bodies, tar-spraying wagons (the two latter types reminders of the numerous Sentinels used by road contractors) also exist.

The Sentinel tar wagons were working in some districts until the late 1950s. At the St Ives depot of Messrs W & J Glossop, the company operated Sentinels until 1958. The last Sentinel to work on road contracts in East Anglia was number 8181 from this depot, which was to be seen at the depot standing disused in 1959. Unfortunately it is not known to have survived. Like other wagon builders Sentinels had an export trade to many parts of the world and doubtless some still survive overseas.

Sentinel standard wagon 3976 was built in 1921 and supplied new to Crosswells Brewery, Cardiff, in September of the same year. Originally on solid tyres and converted to pneumatics by Sentinels in the 1930s, this wagon was working until 1946. Photographed at Sudbury on June 25 1983.

Sentinel number 5407 was built in 1924. The 'Super Wagon' was being cleaned at Woburn rally in 1967. This vehicle was originally a brewers' wagon in Derbyshire.

After a long and hard working life at Brown Bayley Steelworks, Sentinel Standard five-ton tipping wagon number 1716 of 1917, has been carefully restored for preservation. This wagon was in use until 1969. For the last 20 years or so of its working days it was a yard dumper, which involved carrying very heavy loads within the works' area.

A fine example of a Sentinel wagon built in 1926, seen here at the 1984 Stourpaine Bushes rally. This wagon was supplied new to a Dorset Stone Quarry. After its delivery days were over, it ended its working days as a dump vehicle in the quarry.

Above Two very different Sentinel wagons. On the left is 9105, the only 'S8' in preservation. This wagon was built in 1934. On the right is 8571, a 1931-built model 'DG4P', lettered in the name of its last commercial owner, from who it was purchased in 1949 for restoration. This photograph was taken at the Pirton rally, near Hitchin.

Below This superb Sentinel 'DG4' wagon was built in March 1933 and supplied new to The Callow Rock Limestone Co, Cheddar Gorge, who used it for transporting materials to the docks. Photographed while attending the 1983 'Sudbury Mammoth Old Tyme Rallye' in company with several other Sentinels.

Above This fine Sentinel was once part of the extensive steam collection owned by the late Mr T. B. Paisley. Number 8381 was built in August 1930 — a type 'DG4', six-ton with a single chain drive to the rear axle. Photographed June 2 1966 at Holywell.

Below Sentinel 8562 built in July 1931 is an example of the DG6, two-speed six-wheeled type. This wagon was sold new to a haulage contractor in Dudley, Worcestershire and was employed on long distance services for 25 years, including Manchester, Birmingham, Liverpool and London. Later sold and converted to a tar sprayer it ended up in a scrap merchant's yard with the chassis cut in two pieces. Careful restoration has brought the wagon back to the fine condition seen in the photograph taken at Belton in May 1980.

Above One of the few remaining Sentinel 'DG6' wagons, seen here in more recent livery at the 1982 'Expo Steam' event. Works' number 8562 was built in 1931.

Above right 'DG6' seven-ton wagon number 8590 built by Sentinel in November 1931. Supplied new to a cement marketing company at Southampton, this wagon then moved north and saw service as a tar sprayer before being purchased for preservation in 1962. This photograph was taken at Roxton in 1970.

Below right One of the last Sentinel wagons built, number 9293, was completed in May 1937. It is a superb example of the 'S4' type, shaft-driven four-cylinder engines. Photographed while visiting the East of England Showfield for an 'Expo Steam'.

Below Sentinel 'DG6' tipper chassis-cab, number 8351, was built in 1930. This wagon is shown here in the condition in which it was purchased in 1966 for preservation. Amongst its last duties, the wagon was employed in the construction of a new breakwater at Aberdeen harbour—most of this wagon's commercial life was spent in the Aberdeen area.

Above Sentinel 9213 was built in 1936, and is a superb example of the 'S4' model wagons. Note the pneumatic tyres, wipers and electric lighting. This wagon was used by Sentinels as a demonstration wagon for over a year, before being sold to Glendronach Distillery in whose livery it is seen. When restoration was commenced this wagon was just a pile of parts.

Below Sentinel 8992, a type 'S4' built in 1934, was converted to a tanker in 1957, and is often used to help with water problems at rallies. The tank is capable of holding 3,000 gallons of water and weighs 20 tons when loaded. When this photograph was taken the Sentinel was attending a Lincolnshire rally.

Above Sentinel 'S4' number 9151 built in 1934, has been modified to carry a replica steam omnibus body. Photographed standing in the hot sunshine of a July day at the Weeting rally.

Below Sentinel 9277, an 'S4' wagon built in January 1937. This unit is one of the last Sentinels built, being a shaft-driven four cylinder-wagon. Photographed on arrival at the 1984 Stourpaine Bushes 'Great Working of Steam Engines'.

Sentinel 'super' tractor 6504, built in 1926, and matching trailer seen here at the 1983 Sudbury rally. This Sentinel has been restored to a fine condition after being discovered derelict in a scrapyard.

BLACK PRINCE

JOHN R. HARRISON
Contractor
'GARLANDS'
STEEPLE BUMPSTEAD

NT 6347

Left Not many 'S' type six-wheel model Sentinels are in existence. This one is works' number 9103 built in July 1934. In 1981 this wagon attended the Weeting rally, where this photograph was taken.

Right Sentinel timber tractor 8777, a type 'DG4' unit built in 1933 and named *Old Bill*. Supplied new to a wood saw mill, it later passed to an owner in Bucknell, Shropshire, where it was used for timber hauling until the late 1950s. This Sentinel carries a steam winch in the boot at the rear. *Old Bill* had just been unloaded at the 1972 Stourpaine event.

Left Super Sentinel tractor number 7527 which was built in 1928, receives admiring glances from visitors to the 1963 Raynham day as it prepares to move in the parade of engines.

Above right Sentinel timber tractor 8756, a powerful 11-ton engine built in 1933. Relatively few of this design were built, being quite expensive when new. This timber tractor has two 12 NHP engines working from the boiler, one of which is used for propulsion, the other for driving a winch. Photographed in August 1967. This Sentinel is part of the famous Bressingham Collection.

Right The unusual 'Super' Sentinel tractor 5644 built in 1924, *The Elephant*. This five-ton tractor was used for many years to shunt wagons on Teignmouth Quay, Devon. Seen here attending the first 'Expo Steam' event, held at Peterborough in 1973.

An impressive line-up of engines, originating from several builders. The first engine in the line is the veteran Burrell traction engine number 748 built in 1877. Over the last few years the Weeting rally, held near Brandon, Suffolk, has become firmly established as one of the best East Anglian steam events with many of the exhibits performing the sort of duties that they would have done in their working days.

6 The other engine builders

There were other engine-building companies besides those listed in the previous sections. Some built engines in very considerable numbers, while others were comparatively small engineering concerns where total output did not amount to a great many vehicles. This section covers the companies who are known to have survivors capable of moving under their own power. Portables are excluded, as are builders from other countries, although engines built by some of them are now in this country. Examples of American and continental builders' products can certainly be found.

Some of the builders listed here, will have many engines surviving, while others have but one known survivor. The picture selection in this section, as elsewhere, includes the types which are everyday, the rare ones and the solitary survivor — a cross section of all those known to exist.

Wm.Allchin (Northampton) Ltd

This company was located at the 'Globe Works' Northampton. The county of Northamptonshire having large areas of arable land, the company started to produce agricultural traction engines in the late 1870s, producing both single-cylinder and compound designs. In addition they also built road rollers. The oldest surviving Allchin agricultural engine was built in 1890, and the last surviving engine built by the company was produced in 1931. Two ten-ton rollers are also in existence. The other 17 survivors are agricultural engines of single-cylinder and compound designs.

John Allen & Sons (Oxford) Ltd

One 12 NHP ploughing engine built in 1913 represents this company in preservation.

Sir W.G. Armstrong-Whitworth Ltd

All the seven surviving rollers of this manufacturer are compound engines with piston valves.

Atkinson & Co

The company was started in 1907 by two Atkinson brothers and a friend, George Hunt, trading as 'Atkinson & Co, Millwrights, Preston' and operating from premises in Kendal Street, Preston.

With demands for wagons high during the 1914–18 war the company decided to design its own wagon, with the result that the first Atkinson steam wagon was completed in 1916. Number 1 was a six-ton wagon, and was used in the works for several months while performance testing continued. Eventually the wagon was sold locally, remaining with various owners until scrapped in the 1930s. In 1918 the company switched new wagon production to their Frenchwood Works, Preston, Kendal Street carrying out repair work.

During the following years the company built four and eight-tonners in addition to the six-ton design. They also built tractors which were used in Liverpool. In 1923 an articulated trailer unit with a 12-ton capacity was introduced. The company also exported their wagons and fortunately an example of the six-ton 'Colonial' type built in 1918 was discovered in Australia, returning to Britain in 1976. After restoration this unique wagon has become a familiar sight on the rally scene.

Company finances were causing problems in the early 1920s and due to the economic slump, the company merged with Walker Bros (Wigan) Ltd, changing the company name in 1925 to Atkinson Walker Wagons Ltd. Further wagons and rail tractors were built. However times were difficult and the merger only lasted until 1930, when the agreement was terminated. Atkinsons continued for a period of time but the writing was on the wall and the company eventually went bankrupt. The new company formed in 1933 had involvement with steam in the early days but soon turned over to the production of diesel-engined lorries.

Above One of the four surviving Allchin compound traction engines, number 1458, built in 1909. The Allchin is a 6 NHP engine. In the background is a much travelled Burrell road locomotive, 3593, *Duke of Kent*, a 6 NHP three-speed compound built in 1914.

Below left Armstrong Whitworth road roller number 10R2 was first registered in 1923. This 5 NHP compound piston-valve engine weighs ten tons. Only seven Armstrong Whitworth rollers are in preservation.

Below Only two Allchin road rollers remain. Both were owned by Northamptonshire County Council. Pictured here is 1187, which was built at the 'Globe Works' Northampton in 1901. The two surviving rollers are single-cylinder ten-ton models. This rare roller was photographed in 1973, while attending the first 'Expo Steam'.

Above A fine example of the Allchin 6 NHP compound engine, number 1458 was built in 1909. Not many Allchins have survived. This engine spent a lot of its working days in Gloucestershire. Originally designed for road haulage and agricultural use, it is sprung front and back, which together with the 'rubbers' would improve the ride considerably.

Above Atkinsons built this interesting six-ton double-crank compound 'Colonial' type wagon in 1918. It is thought to have been supplied new to a brewery at Perth Australia. Later employed at a gold mine for some time, it was eventually abandoned in the bush for a great many years.

Three veteran wagons which have made the long journey to Australia. Having been discovered derelict and recovered for restoration they each made the return journey to be carefully restored to the condition which can be seen in this photograph. On the left is the Aveling & Porter wagon 9282 of 1921. The right hand engine is Foden 4086, now in Germany, while the middle wagon is six-ton Atkinson number 72 built in 1918.

Babcock & Wilcox Ltd
Five road rollers by this company survive, carrying manufacturers' plates bearing the 'Babcock & Wilcox' name. In fact the engines were built by Clayton & Shuttleworth Ltd, a well known engine builder taken over by Messrs Babcock & Wilcox Ltd in 1924. After applying their own plates to the Clayton & Shuttleworth rollers, they were sold in 1926.

Brown & May Ltd
Survivors of the engines built by the Devizes, Wiltshire, company are mostly portables, with the exception of a single road roller and a 5 NHP showman's tractor, works number 8742 built in 1916, and named *General Buller*. This compound two-speed engine was supplied new to a North Wales amusement caterer, but, like so many road engines ended her working days running a set of threshing tackle. After the threshing business closed the engine lay out of use for a number of years.

Clayton & Shuttleworth Ltd
Another of the old established Lincoln-based engine builders, Clayton & Shuttleworth produced engines until 1930 when the company was taken over by another Lincoln firm, Marshall & Sons Ltd.

The Clayton engines in preservation are mostly agricultural or road rollers, although other interesting engines exist, such as the sole surviving tractor, number 49008 built in 1934. This is a five-ton 4 NHP compound engine, and can now be seen in East Anglia. Clayton & Shuttleworth wagons are represented by four survivors, all are five-ton wagons built in 1919 and 1920. Three are preserved in Britain, and one in Ireland.

Clayton Wagons Ltd
One wagon built by this company is with us today, being part of the famous Thursford collection. The wagon was built at Lincoln in 1926 and is a six-ton three-speed wagon with a three-way tipping body and Ackerman-type steering. Originally it was owned by Norwich corporation, ending its working life as a tar sprayer.

Above A very rare engine is Brown & May 5 NHP showman's tractor *General Buller*. This two-speed engine was built in 1916 to the order of a North Wales amusement caterer, ending its showland days in the Nottingham area. Like so many road engines General Buller ended its working days on threshing work, until restored by its owner for preservation.

Left Clayton & Shuttleworth (nearest camera) and three Marshall agricultural engines, caught in a quiet moment as they simmer gently in the spring sunshine.

Below Another rare wagon, which can be seen on the rally field from time to time, is this fine Clayton & Shuttleworth, makers' number 48347 built in 1919. This five-ton compound wagon is preserved in the county in which it was built many years ago.

Opposite Another rare wagon is *Fenland Princess*, a Clayton & Shuttleworth 4 NHP built in 1920 as works' number 48510, this five-ton wagon was purchased for preservation in a derelict condition. Following many hours of dedicated work, the engine made its first public appearance in 1976 at the Haddenham rally.

Above Another view of the 1920 Clayton & Shuttleworth 4 NHP five-ton wagon *Fenland Princess* photographed at Roxton in 1976, the first year it returned to steam.

Below Early rally days. Clayton & Shuttleworth 7 NHP, 48224, was built in 1919 and is seen here in fine form as it proceeds slowly round the main ring at one of the early 1960s Ickleton rallies.

Left Belton Hall was the setting for several rallies held in the late 1970s and early 1980s. Standing in delightful grounds is Clayton & Shuttleworth, works' number 46817 built in 1914. This convertible engine spent its working days as a roller. Among its owners were the War Department and Messrs Buncombes of Highbridge. On being purchased for preservation, in a very derelict condition, the engine was converted to a traction engine.

Right Photographed in September 1962 at the Ickleton rally, Clayton & Shuttleworth agricultural 48224, a 7 NHP single-cylinder engine had been completely rebuilt in the preceeding years.

Right This Clayton & Shuttleworth ten-ton roller left the works in 1923, being supplied new to Messrs Buncombe of Highbridge, Somerset. The roller has not long been in preservation and was in fine condition when attending the 1984 Stourpaine event.

Right Not very many Clayton & Shuttleworth road rollers have survived. Here works' number 48971, a single ten-ton model was photographed while on display at the 1980 Rempstone rally. This engine was built at Lincoln in 1925. Note the plates on the rear rollers to remove stones etc.

Left Awaiting its turn for restoration in 1966 was this Clayton & Shuttleworth agricultural engine, 43200, of 1910. This single-cylinder engine is a 7 NHP design. Among the missing parts was the smokebox door.

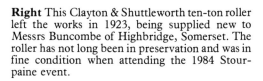

One of the rare Davey Paxman agricultural engines. This one, number 13073, is the oldest known survivor, being built in 1907. This 7 NHP single cylinder two-speed engine, was at one time part of the collection of Mr Paisley. Much of the engine's working life was spent in Hertfordshire, including a period of time for the Hertfordshire County Council. The engine also attended the centenary celebrations at the Paxman works in 1966.

Davey Paxman & Co Ltd

Four agricultural engines are with us today from this Colchester company, which celebrated its centenary in 1966. On that occasion one of the engines surviving, number 13073 built in 1907 (and incidently, the oldest survivor of the agricultural engines) was exhibited at the works. At that time it was in the ownership of the late Mr Paisley.

Wm. Foster & Co Ltd

Another of the Lincoln based and old established engine builders, this company was especially renowned for their agricultural and road locomotives. Approximately 26 general purpose, and seven showman's road locomotives are with us today.

In addition several 4 NHP tractors are located in various parts of the country. A roller and several portables also exist. The final number of Fosters in preservation is made up by eight showman's tractors of (mostly) 4 NHP.

An extremely rare Foster 'overtype' wagon has returned from Australia, and is in the course of restoration. Considerable numbers of these wagons were built, some having tipping bodies. The one refered to is the only known example in this country.

The famous 'Wellington' tractor built by this company was introduced in 1904 to comply with the road traffic acts of that year.

Foster 7 NHP agricultural engine 3682 of 1908 vintage—caught by the camera while attending the 1982 Haddenham rally.

Part of the famous line-up of showman's engines for which the Stourpaine event is famous. In this 1984 picture the engine on the right hand side is Foster 14564, *Victoria*. The engine in the centre is Garrett 33986, *Queen of Great Britain*, while in the background several of the veteran fairground rides can be seen in full swing.

Above Early preservation days. Foster showman's 14153 was built in 1916 as a road haulage engine. Later it was returned to the makers and converted to a showman's, being purchased by the Thurston family. In the photograph *Admiral Beatty* is seen before it was restored to full showman's glory, having been employed on agricultural duties. Note that the dynamo has been removed, together with some of the brasswork. The roof too had seen better days — Chatteris rally July 28 1963.

Below A fine example of a Foster general purpose engine, caught by the camera at the 1973 Carrington Park rally, 2373 is an 8 NHP single-cylinder agricultural engine built in 1899.

Fowell & Co Ltd

In 1876 the 'Cromwell Works' at St Ives, Huntingdonshire commenced business in premises not far from the railway station. In those days St Ives was a small market town in the heart of an agricultural area with an important cattle market and good rail connections via Cambridge, March and Kettering. The nature of the area provided considerable work for the company, particularly in repairs to traction engines and other boilerwork.

New construction of engines was carried out somewhat erratically over the years and the total number built was not very large. The works' list, included, apart from traction engines, portables, and tug boat engines, a boiler and even an oil engine. The last engine to be completed was number 109. Another, number 110, was started prior to the 1914–18 war but never completed.

Seven Fowells survive, three of which are, at the time of writing, within a few miles of St Ives, one other example is in Ireland. The remaining three are in various parts of Britain.

One of the survivors is 108, an 8 NHP engine, the last-but-one engine to be constructed, being completed in July 1922. At one stage three of the remaining Fowells were in the collection of the late Mr T.B. Paisley at Holywell, being dispersed widely at the engine sale held there in 1980. Unfortunately several other Fowells, still intact in the 1950s, fell into the hands of the scrap merchants.

One of the two remaining Fowell 8 NHP engines, works' number 108, was built in 1922. Only seven engines built by this St Ives company have survived into preservation.

Fowell number 92 built in 1903, seen here attending the 1963 'Raynham Day'. The event held in 1962–63 attracted a large number of entrants. The engine is a 7 NHP, now carrying the name *Roundhead*. It was formerly part of the collection of three remaining Fowells owned at one time by Mr Paisley.

St Ives built, Fowell number 108, was built in 1922. In the close up view of the front end of this engine the high standard of restoration can be clearly seen. Over 100 engines were built by this small firm but only seven survive today.

An interesting rear-end view of Fowell number 97, built at St Ives in 1907, now carrying the name *The Black Prince*. Number 97 is one of the 7 NHP design, of which five engines have survived. Formerly in the Holywell Collection, she moved south in 1980, and was photographed in the working enclosure at the 1984 Stourpaine event.

A low-angle shot of Fowell number 93, *The Abbott*, a 7 NHP engine built at the 'Cromwell Works', St Ives. The position of the front wheels on the Fowells, can be clearly seen in this photograph taken at the 1979 Haddenham rally.

Fowell number 91 is the oldest of the seven surviving engines built by the St Ives manufacturer. This single-cylinder 7 NHP engine was built in August 1902. Note how the front wheels are set back under the boiler.

Richard Garrett & Sons Ltd

Another East Anglian builder was Garretts of Leiston, Suffolk. Quite a considerable number of engines built by this company are in existence today, the majority being tractors, or showman's engines. Tractors were built in large numbers, of conventional design, but inclined to be heavier than several of the competing companies' designs. Garretts did produce most of the other types of agricultural and road engines and examples of traction, roller, and portable are all with us today, as are three wagons built by them. One 6 NHP showman's built in 1908 is preserved.

A unique and unconventional design produced by the company in an attempt to counter the increasing number of internal combustion tractors was the 'Suffolk Punch'. The design was very manoeuvrable having chain drive and controls similiar to a motor car. While the design was well thought out, it did not catch on and only a small number were sold. Fortunately, one survives today, number 33180 built in 1919. This interesting sole survivor attends rallies over a wide area, attracting considerable interest.

Several years later Garretts introduced an updated 'Suffolk Punch'. This model was even more unconventional in design, having chain drive to both of the rear axles, thus giving the vehicle a four-wheeled drive. None of this later type are known to have survived.

As mentioned elsewhere the last Burrell engines were built by Garretts, the Burrell rights having been taken over by the Leiston company. Among the last Burrells built was 4092 *Simplicity*, the last showman's engine to be built, and unfortunately scrapped.

This Garrett single-cylinder 7 NHP agricultural engine was built at Leiston in 1916 and supplied new to a Great Yeldham owner. It was commandered by the War Department and shipped out to France. On its return it spent the rest of its working days in Essex. This particular engine has spent a great many years attending rallies, in fact since the mid 1950s.

Above The unique 5 NHP Garrett 'Suffolk Punch' tractor *The Joker*, number 33180 was built in 1919. The design was introduced to counter the steadily increasing numbers of internal combustion-engined tractors. Although well thought out, the design was not popular and very few were built.

Below Two Garrett eight-ton 'QL' Class wagons are in preservation. This design is a most interesting undertype wagon, very advanced for its day. The wagon is fitted with Ackermann steering, the drive is taken by geared drive and differential and finally by chains to the rear axle and the rear wheels have covers giving a solid appearance. Built in 1931 as works' number 35465.

Left Quite a number of 4 NHP Garrett tractors have been preserved. This one is number 33782, built in 1919, for a saw mill in Berkshire. The engine now carries the name *Dorothy*, and was photographed at the 1984 Haddenham rally.

Right This photograph shows how engines can change in a fairly short period of time. Here, when photographed at Great Wymondley in 1965 Garrett 33380 was a tractor. The engine is now a showman's tractor.

Right Garrett showman's tractor 33380 was built in 1918 for the Ministry of Munitions, it later worked for Coulsons of Park Royal, London. It was purchased for preservation and later converted to showman's specifications. *Saphire* is a two-speed compound 4 NHP engine weighing five and a half tons. This engine is shown as a tractor in an earlier 1965 photograph.

Right Garrett 4 NHP showman's tractor *Lord George* was built in 1918, as works' number 33358, and supplied to the War Department. After its government service, it was sold and converted to a showman's tractor, being used until 1938, when it spent the next 15 years lying derelict. As can be seen, the engine is in the course of receiving a complete overhaul.

Left Late September sunshine picks out the detail on this Garrett single-cylinder agricultural engine built in 1916. Apart from service in France during the 1914–18 war, the engine spent its working life in the Suffolk area. It has attended rallies for a number of years now.

Above Many years of this 4 NHP Garrett tractor's working life were spent as a showman's tractor. Number 33278, *Princess Mary*, was built in May 1918 and supplied new to the War Department. In 1922 it changed hands, and was converted to a showman's. After the last war the engine again changed hands and was converted back to a haulage tractor. The engine retains more brasswork than is usual — a reminder of its showland days.

Below The traction engine rallies held at Great Wymondley near Hitchin in the 1960s attracted an average entry of 35 to 40. Here two youthful admirers look over Garrett five and a half-ton 4 NHP showman's tractor, works' number 33941. This engine is named *Evening Star* and was built in 1920. Photographed on June 6 1965.

Above Originally built as a road roller, and owned by Shropshire County Council, this Garrett 4 NHP was converted to a showman's tractor in 1968 and is now named *Yeovil Town*. Works' number 31193 was built in May 1913. Photographed here at the 1984 Stourpaine Bushes event.

Below Banbury steam engine rally held on June 28 and 29 1969, attracted several showman's engines. Among them was this Garrett showman's tractor 33505, *Rambler*. For many years the engine travelled for a Birmingham showman. During the war it was used on agricultural work but it was later returned to showman's condition.

T. Green & Sons Ltd
Only five road rollers built by Messrs T Green & Sons Ltd, 'Smithfield Iron Works' Leeds survive in preservation today. The five survivors are single-cylinder, slide-valve engines, two of eight-tons each, one of nine tons and two ten-ton models.

Gibbons & Robinson
One 1891-built agricultural engine of 8 NHP survives.

Richard Hornsby & Sons
Most of the survivors built by this company are portables, with the exception of one 8 NHP agricultural engine built in 1890. This company later merged with Ruston, Proctor & Co Ltd.

W. Lampitt & Co
This company too is represented by only one solitary survivor, an agricultural engine of 1887.

J. & F. Howard, Bedford
The 'Brittania Iron Works', Bedford, had no surviving examples of its products in Britain until 1982, when engine 201 returned home, having lain derelict in Australia with many parts missing, for many years. This interesting engine, built originally in 1872, has been superbly restored and attended its first post-restoration event in Britain at 'Expo 83'. Number 201 amply demonstrates both the craftsmanship of her original builders and the dedicated skills of her restorers. Only one other Howard is known to exist, and that is in the United States.

Left Garrett showman's tractor 33566 *Little Billy*, a 4 NHP compound built in 1919, photographed at the 1983 Sudbury 'Mammoth Old Tyme Rally' held at Melford Hall on June 25 and 26.

Below left Another view of the 1984 Stourpaine showman's line up, in this picture Garrett *Queen of Great Britain* and Aveling & Porter *Princess Victoria* were caught by the camera in a rare few moments that they did not have a crowd of admirers round them.

Below Britannia Iron Works, Bedford, was the birthplace, way back in 1872, of this unique Howard traction engine which spent its working life in Australia. Discovered derelict, it was returned to this country, and restored to the superb condition which can be seen from the photograph. 201 was last thought to have worked about 1908 on a shearing plant and is the makers' number as far as is known — no brasswork remained on the engine, and unfortunately much of Howards' records have been destroyed.

Leyland Motors Ltd

Starting from small beginnings at Leyland, Lancashire, the Leyland steam vehicles were produced up to 1926, when the steam spares were transferred to Messrs Atkinsons, Preston (steam wagon builders in their own right). From this date responsibility for Leyland-built vehicles, spares, repairs, etc, was taken over by Atkinsons.

Early developments were on steam vans, steam buses and wagons, eventually building up to a production run of about 100 each year in a wide range. However development of internal combustion engine vehicles was also proceeding and with the outbreak of the 1914–18 war, contracts were placed by the War Department for petrol vehicles. This created a space problem and the company negotiated the purchase of an empty mill at Chorley, intending steam vehicle production at this site. In the event steam vehicle production was stopped until after the war.

In 1919 the company Leyland Motors Ltd was formed and steam vehicle production was again in full swing. In the 1920s the sale of steam wagons began to decline and the range on offer was reduced, until in 1925 steam vehicles were available on early delivery. Finally a decision was made to cease steam vehicle production altogether and the transfer of spares etc, to Atkinsons took place. Leyland Motors Ltd concentrated on internal combustion vehicles.

Over the years the company had produced a considerable number of wagons, a great many of which had been exported. The sole surviving Leyland wagon was found derelict in Northern Queensland, Australia, probably having spent all its life around the logging camps in the area. The wagon was crated and returned to Britain where restoration work was commenced by works' apprentices in early 1968. After two years of careful and painstaking restoration this interesting wagon was restored to the condition in which it is seen today on its occasional rally and vintage vehicle event attendances.

Mann Patent Steam Cart & Wagon Co Ltd

This is another company whose works were in the county of Yorkshire, Manns works being situated in the Hunslet district of Leeds.

Few engines built by the company have survived. The majority of the 12 survivors are tractors but there is one road roller and two are wagons.

The Mann steam cart, a rather unusual tractor design, was intended for operation by one man. The firebox was on the offside, with the firehole door sited so that the driver could look after the fire. The driving position was the only crew space on the engine. These tractors were designed for direct ploughing but could also be used with a cart body fitted.

Above The sole surviving Leyland wagon built in 1914, found derelict in Northern Queensland, Australia. It was shipped back to Britain in a crate and carefully restored by Leyland Motors Ltd apprentices over a two year period. Photographed at one of its attendances at the 'Expo Steam' event held on the East of England Showfield, Peterborough.

Above left A second view of the Howard, this time from the opposite side. Both of these photographs were taken at the 1983 'Expo', which was the first event that this unique engine attended after restoration. A short time later the engine was in steam outside the works at Bedford that it left well over a hundred years ago. One other Howard is known to exist — this engine being in America and in need of restoration.

Below In the 1960s and 70s this neat little Mann 4 NHP tractor, *Little Jim*, was frequently seen at rallies in East Anglia. Built in 1920, works' number 1425, this design was used for ploughing and other agricultural duties.

Above Only two Mann wagons are thought to be in existence. This one, works' number 1120, is the oldest of the two, being built in 1916. It is a compound five-ton wagon, built by Mann Patent Steam Cart & Wagon Co Ltd, Leeds.

Above right A fine example of the Marshall traction engine, number 83780, built in September 1928. This single cylinder ten and a half-ton engine is a 7 NHP design. The Marshall badge can be clearly seen on the side of the boiler. Note the difference between this 1963 rally scene, and the present day.

Right Early rally days — the second Ickleton rally was held on September 8 1962. Here Marshall agricultural engine 79342, built in 1926, stands in front of other Marshall equipment. The veteran is a fine example of a Marshall compound. Mounted on springs, it is a 6 NHP ten-ton engine.

Marshall, Sons & Co Ltd

The company began in Gainsborough Lincolnshire in the 1840s, commencing traction engines construction in 1876, building up to become a large concern and eventually taking over another established Lincoln company, Clayton & Shuttleworth in 1930.

Over the years Marshalls built up an excellent export market, with their engines travelling far and wide throughout the world. Several hundred Marshalls survive, a great many of which are to their agricultural engine designs. Considerable numbers of these were built. One of the oldest agricultural engines in preservation is a Marshall of 1886, with many others of the late 1880s and 1890s also preserved. Numerous road rollers survive, ranging from a 1900 engine, through to rollers built in 1943–44.

The company was extremely well known for its portables. Some of the earliest survivors date back to 1874 while the most recent are units built in the 1940s. The excellent engines built by Marshalls are found throughout the country, and engines of their manufacture are represented at most engine events.

Above This interesting line-up was at the end of season 'Steam-up' held in October 1984. Engines in the line include Marshall & Aveling tractors, together with a Wallis & Steevens general purpose engine. At the end of the line are two Fowlers.

Above left An engine typical of the Marshall agricultural designs. This one, works' number 57375, was built in 1911 as a single-cylinder slide-valve engine of 7 NHP weighing ten and a half tons.

Left Marshall 7 NHP agricultural engine, number 61880, was exhibited at the 1913 Royal Show together with a set of threshing tackle. Its first owner was in Hertfordshire, where it worked for many years until moving north to the Grantham area.

Right Ready to provide power for the threshing demonstration at Marsh Gibbon. Marshall general purpose engine 52367, a single-cylinder 7 NHP unit built in 1909. In recent years demonstrations of threshing have proved popular with the general public, who find it fascinating to watch the old methods, now easily replaced by the modern equipment for grain harvesting.

This Marshall traction engine, 74871, left the Gainsborough works in 1921. A neat compound slide-valve engine, it is a convertible type 5 NHP engine weighing ten tons.

Marshall 'Q' Class ten-ton road roller, works' number 76796, was built in 1923. This roller is one of four consecutively numbered engines to have survived, all being the same weight and 'Q' Class design.

Marshall road roller, 76116, photographed at Rempstone rally in 1980. The roller is lettered Goole Rural District Council. The engine is a 'S' Class compound piston valve, weighing eight tons.

J. & H. McLaren Ltd

This company has been proud of its well-deserved reputation since its start in 1877. The 'Airedale Foundry' was the location of another of the famous Leeds engine builders.

Surviving engines built by McLarens are mainly agricultural designs, although two ploughing engines and two showman's road locomotives exist. The larger of the showman's is *Goliath* number 1623. This engine was supplied to the War Department as a haulage engine, and converted to a showman's for Pat Collins amusements. *Goliath* was used in company with two other similar engines *Samson* and *The Whale*. All three travelled widely until the mid 1950s but unfortunately *Goliath* is the only one of the trio to survive. It is the last remaining 10 NHP McLaren showman's, the other showman's being an 8 NHP engine in Ireland.

Four compound road locomotives represent the haulage types built by the company. One is a 5 NHP engine, one an 8 NHP, number 1421 *Captain Scott* built in 1913, and one of the two remaining 10 NHP engines, 1652 *Boadicea*, is equally well known.

Of the three tractors in preservation, number 1837, one of the last engines built by McLarens, and named *Bluebell*, is very well known as it regularly attends rallies under its own steam, travelling considerable distances. On more than one occasion *Bluebell* has even entered the London to Brighton run. The survivors are completed by a solitary road roller of 1908 vintage.

Most the the surviving ploughing engines are Fowlers, but at least two ploughing engines built by McLarens of Leeds are also preserved. This photograph is of 1541, built in 1918, a compound 12 NHP engine, weighing 14 tons.

Above The big McLaren 14-ton road locomotive *Boadicea* was often seen on the highways during the rally season in the mid sixties. Here number 1652 built in January 1919 makes its way across the 1963 Chatteris rally field.

Right Veteran McLaren agricultural engine, number 127, was built at the Leeds works in 1882. *The Little Wonder* is a single-cylinder 8 NHP weighing eight tons. Note the various items of equipment being carried between the front wheels.

Below right This McLaren traction engine was supplied new to an owner in Tipperary, Ireland. After passing through several hands, it ended up derelict, like so many others. Number 435 of 1892 here drives slowly round the field at the 1978 Roxton rally.

Below Early hours at the 1980 Rempstone rally, and McLaren 1421, *Captain Scott,* has already been coupled up to a Ransome threshing drum, ready to give demonstrations throughout the day. Soon this 'Mac', built in 1913, would be providing power smoothly amidst a crowd of interested onlookers.

Photographed at Stilton on its way to an event, McLaren 4 NHP number 1837, built in 1936. This nippy five-ton tractor travels mostly under its own steam, it has also entered in the London–Brighton commercial vehicle run. The 'Mac' was owned for several years by Watson & Haig of Andover who used it for hauling timber, it also spent some years working at Southampton docks. Bought in working order from Pocklington, Lincolnshire, the engine was driven to its new home, having been in preservation for many years.

Ransomes, Sims & Jefferies Ltd

Originally this company traded as Ransomes, Sims and Head. In 1881 it became Ransomes, Head and Jefferies and then in 1884 it again changed, this time to the above title. This company is one of the East Anglian builders with its 'Orwell Works' located at Ipswich, Suffolk. Considerable numbers of engines built by the company are in existence today. Most are agricultural engines and tractors but a considerable number of portables, built over a long period of time, also survive. The company did try to break into the market for showman's engines, but their showman's engines did not prove very popular, and as a result very few were built. On the other hand they did produce an excellent 4 NHP light tractor, of which design several survivors are in preservation.

Robey & Co Ltd

This is a Lincoln company with a long history of engine building. Types included a great many portables and boilers of many designs and sizes, which have been exported all over the world.

The first agricultural engine built by the company appeared in 1862 and development continued until after the turn of the century when production ceased. Production restarted again about 1905 and continued until the late 1920s.

Wagons have only one single survivor, a six-ton overtype unit built in 1925. The earlier 1907 undertype, with vertical boiler does not survive.

Robeys are well known for their tandem road rollers. These are double or tri-tandem units, and 12 survivors of the various types still exist. Robeys also built a tractor which they named the 'express' tractor. Nine were built but it is not known how many still exist

Above On a rather grey day at Carrington Park, Lincolnshire, some of the agricultural engines raise steam. First in line is Robey 42675. Standing next to the Robey is a Ransomes Sims and Jefferies, while in the background stand engines built by other well known makers.

Below This fine example of a Ransomes Sims & Jefferies traction engine was built at Ipswich in 1918. The engine weighs nine tons. Most its working life was in Staffordshire before being finally retired in 1943 with another Ransome engine bearing a preceeding registration number.

Left This neat little Ransomes Sims & Jefferies 4 NHP tractor, was built in 1911. The engine carries the name *Backus Boy*, and was photographed at Raynham in 1962. The tractor was used as a demonstration engine by the manufacturers, attending shows over a wide area.

Right Robey wagon 42567 was built in 1925, a six-ton compound three-speed wagon. Photographed on May 28 1973 at the tenth annual rally to be held at Carrington Park. Note how the front wheels are set well back on the wagon.

Left Many examples of engines performing the tasks for which they were designed, can be seen at Stourpaine. Here Ransome Sims & Jefferies 7 NHP, 26995, is coupled to a wood sawing bench.

Right Robey five-ton tractor, number 41492, was built at Lincoln in 1923 and supplied new to the RASC as a haulage engine, conversion to showman's type taking place much later.

Left Only nine Robey 'Express Tractors' were built, of which only two or three are thought to have survived. Pictured here is number 43388, a 4 NHP compound piston-valve unit weighing seven tons and built in 1929.

Robinson & Auden Ltd

This company is represented by six surviving examples of its products, five of which are portable engines and only one of which is a traction engine. The traction engine was at one time in the collection of Mr Paisley of Holywell. The surviving traction engine is works number 1376 built in July 1900, an eight-ton two-speed 6 NHP general purpose engine which was purchased derelict and restored in the Holywell workshops. Many parts were made and machined to replace missing items.

This extremely rare engine was sold in the engine sale of October 1980 disposing of Mr Paisley's collection.

Ruston, Proctor & Co Ltd/ Ruston & Hornsby Ltd

Another of the Lincoln-based engine builders, the 'Sheaf Ironworks' was the base of Ruston, Proctor & Co Ltd until the company was succeeded by Ruston Hornsby Ltd in 1919.

Of the engines originally built by Ruston, Proctor & Co Ltd, the survivors are mainly agricultural and portable engines, but five road rollers and one five-ton 4 NHP tractor also exist.

After the company was taken over — to become Ruston & Hornsby Ltd — they continued to produce portable engines, many of which survive. Surviving agricultural engines and tractors are represented by 14 preserved vehicles, along with a few road rollers.

Left Robey 44083 is a rare example of the tri-tandem road roller produced by the company — only one other tri-tandem exists. This roller was completed in May 1930. For many years it was owned by Wirksworth Quarries Ltd, working in many parts of the country including Suffolk, Bedfordshire and Lincolnshire. The roller was in commercial use until the late 1960s.
Below left Robey tri-tandem roller, makers' number 45655, was built in 1930. Here the roller is seen at the Great Wymondley rally, held near Hitchin on May 29 1966.
Below A photograph of an engine under restoration at Holywell, this time the engine is the unique eight-ton two-speed Robinson & Auden traction engine built at Wantage in 1900, makers' number 1376. This engine was acquired in 1965, and during restoration a new firebox was fitted in the workshop.

Above The unique Robinson & Auden 6 NHP single-cylinder engine, as restored, shortly before attending the 1976 'Expo Steam'.

Below Ruston & Proctor 50278, shortly after receiving a major overhaul. Seen here at Chrishall Grange on September 30 1984. This engine made a national record for hay baling while requisitioned by the War Department.

Ruston & Proctor number 33471, of 1907
vintage, photographed in heavy conditions at
the 1967 Woburn Park rally. The engine is a
Class 'SH' single cylinder of 7 NHP, weighing
in at 9½ tons.

W. Tasker & Sons Ltd

In 1809 Robert Tasker set up business as a blacksmith at Abbotts Ann, a small village in Hampshire. Four years later he opened the famous 'Waterloo Ironworks' in Andover.

One of the designs for which the company is well known is the 'Little Giant Tractor'. One 'Little Giant' was kept in a shed near Crystal Palace London and was used by the RSPCA to help horses pull loads up steep hills in the district. The majority of the surviving Taskers are tractors to various designs but a sole surviving wagon, built in 1924, is also preserved. Another 'one off' survivor, is the showman's tractor, number 1822, built in 1920, and now resident in the eastern counties. The remainder of the surviving engines include a few agricultural vehicles and road rollers. The last engine built by the company was a ten-ton road roller in 1926.

Several of the company's products, covering most designs, are held in preservation by the 'Tasker Trust'. These include early stationary boilers, an 1893 agricultural engine and the sole surviving wagon.

Left This Ruston Proctor 4 NHP tractor, 52607, was built in 1918 for the War Department, being part of a large order. In 1921 the engine moved to an owner in Coventry, where it was used for coal haulage. Later owners used the engine on agricultural duties. In the photograph the Ruston was coupled to another engine, having hauled a heavy load of timber up the incline at Stourpaine.
Below left An engine which has attended many rallies under its own steam is *The Muddler*, Ruston Proctor 8 NHP Class 'SH' traction engine, works' number 35501 of 1908 weighing ten and a quarter tons.
Below Tasker showman's tractor 1822 built in 1920. This 4 NHP type 'B2' tractor was converted to showman's condition when approximately three years old. It is thought to be the sole surviving Tasker showman's tractor. Photographed on a sunny July day at the 'Weeting Steam Engine Rally', held near Brandon, Suffolk.

Left Tasker 'B2 Little Giant' tractor, works' number 1895. This compound, three-speed chain-drive 5 NHP tractor was built in 1922. The engine spent its working life in Scotland.

Left Carrying the name *The Horses Friend*. Tasker 4 NHP Class 'A1' tractor, number 1296 was built in 1903. The letters 'RSPCA' can be seen on the motion cover. This tractor was used to assist horses hauling heavy loads up steep hills. Photographed at the 1972 Stourpaine Bushes rally.

Right One of the two remaining Thorneycroft wagons, this one built in 1902 as works number 115. Seen here while attending the 1981 'Expo' event. The wagon carries the lettering 'County Borough of Bournemouth', and on the side, 'Speed five miles per hour'.

John L. Thorneycroft & Co Ltd

Two veteran wagons built by this company survive as does an early steam van built in 1896. The two wagons are number 39, fitted with a brewer's dray type body, and number 115, built in 1902, which is a very spartan flat platform unit without a cab.

Wallis & Steevens Ltd

Within the ranks of preserved engines, quite a considerable number built by this Hampshire company have survived. Probably the most widely known are the 'Expansion' traction engine and the 'Advance' road roller designs. The company was started in 1847 by a Mr A. Wallis as a general engineering company, manufacturing castings and farm implements. Not long after the company was formed it started building portables and small stationary engines.

In 1877 the first traction engine was built. Known as the 'T' series, it was an 8 NHP two-speed engine. Later developments of the design were 6 and 7 NHP engines. The first of many road rollers, an 'Expansion' type engine, was built by the company in 1890. Later introductions were the three-ton tractors, compound designs in the range 5 to 8 NHP. Another development was the oil-bath compound engine. The company also built wagons and a five-ton example survives.

The year 1923, saw the introduction of the prototype 'Advance' road roller, a general design with high pressure cylinders and piston valves which was to continue in production for a considerable number of years and include various weight models. The company exported to many countries. In 1925, the three-ton 'Simplicity' roller was developed. Road rollers powered by other fuels were also built.

One original showman's tractor built in 1914 survives with the original twisted brasswork. The majority of the company's designs in preservation are road rollers with strong representation of the 'Advance' design, including the last Wallis and Steevens engine built for the home market in 1939. The original prototype 'Advance' roller built in 1923 has also, fortunately, survived. Examples of traction, tractor and 'Simplicity' rollers can be found, together with the one five-ton wagon.

Development of Basingstoke saw the old works completely razed to the ground, and a new factory located a short distance away.

Above On the early wagons the crew had very little protection from the elements, as can be seen in this low-angle shot of Thorneycroft Number 39 built in 1901. This wagon is a five-ton 5 NHP vehicle fitted with a brewers' dray body. *Dorothy* is a regular entrant at rallies held in eastern England.

Below Ready to perform its duties in the threshing area of the 1984 Stourpaine '16th Great Working of Steam Engines', Wallis & Steevens number 7102, a 7 NHP single-cylinder general purpose engine built in 1909. Most of this engine's working days were spent in the south of England.

Above One of the last events held each year is the 'Vintage Ploughing and Farm Machinery Show' held on September 29 and 30 1984 at Chrishall Grange on the Cambridgeshire–Hertfordshire border. Several steam engines were present including Wallis and Steevens ten-ton 'Advance' roller number 8096 built in 1935 for Barry Council. This roller has had a complete overhaul recently and is now in excellent condition.

Below Wallis & Steevens 3 NHP tractor, *Goliath* was built in 1902. This engine is an example of the early three-ton tractors built by this company. Little history is known of this engine except that in its later working days it worked in the Reading area for a brick and tile works.

Left This 4 NHP six-ton 'Advance' roller was built by Wallis and Steevens in December 1930. Here it moves slowly round the field at the Roxton rally.

Right One of the unusual five-ton 'Simplicity' road rollers built by Wallis & Steevens. Only six of these rollers are known to remain in the country.

Right This close-up view of the ten-ton Wallis & Steevens 'Advance' roller, shows in detail the two high-pressure cylinder and piston valves.

Right Wallis & Steevens oil-bath compound tractor of 4 NHP was built at Basingstoke in 1926. It attended the Raynham rallies held in 1962–63 where it was photographed in 1962.

Left An unusual example of a Wallis & Steevens 'Advance' road roller, number 8100 was built in April 1936. All 'Advance' rollers have two high-pressure cylinders and piston valves. This roller is one of the three surviving eight-ton models, the other two being in South Africa. This engine was supplied new to Hampshire County Council.

For a number of years this sturdy Wallis & Steevens five-ton oil-bath tractor, was employed on wood cutting, for which a larger flywheel was fitted. Originally the engine was supplied to an owner in Dorset. In 1972, number 7289, built in 1912, was photographed at a rally held at Duxford Grange.

Index